ELECTRONICS UNRAVELED-
A New Commonsense Approach

ELECTRONICS UNRAVELED-
A New Commonsense Approach

by James Kyle

FIRST EDITION

FIRST PRINTING—JUNE 1974

Printed in the United States
of America

Hardbound Edition: International Standard Book No. 0-8306-4691-4

Paperbound Edition: International Standard Book No. 0-8306-3691-9

Library of Congress Card Number: 74-79583

Contents

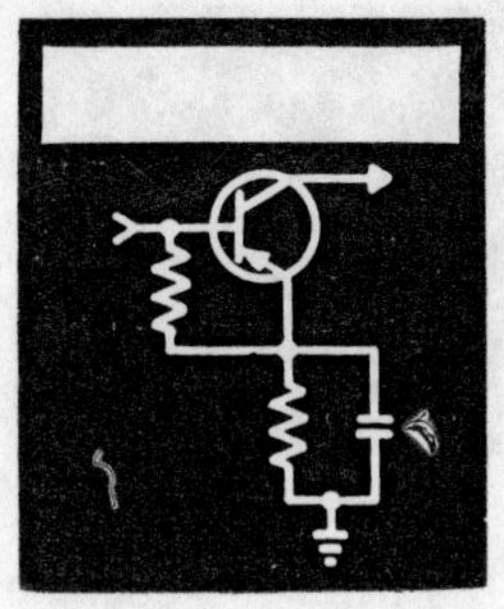

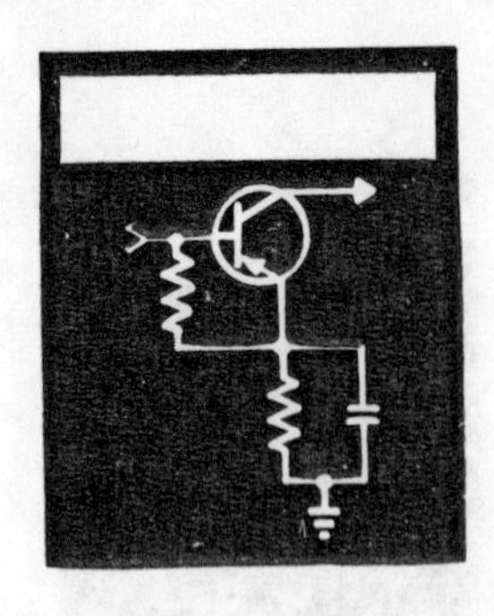

Foreword

It's been 15 years now since a hungry young police reporter who also happened to be a ham discovered that he could add to his income by writing "how-to" articles for the radio magazines.

In that 15 years, I have turned out several reams of copy—most of it dealing with the most basic elements of practical electronics. It always seemed to me as if there were almost a conspiracy to make this material seem difficult—and it really isn't all that hard to understand.

Along the way, I developed a number of unusual ways to present the material, which I think help show just how simple the art of electronics is at bottom. You'll find them in this book, as well as in a few others with which I have been associated, but you're not very likely to find them elsewhere. Instead, you'll be confronted with an avalanche of jargon (a word which I have defined as "a special language designed to impress the outsider with how much the user knows"), which reinforces the ideas most people have that anything as miraculous as electronics just **has** to be complicated.

I am told that anyone who really knows what he's talking about can afford to use words anyone can understand. I hope I've used words anyone can understand. More than that, I hope I've made it interesting enough to keep you with it, until you too see this fascinating subject in its true simplicity and (even as I did many years ago) come to appreciate its awesome possibilities.

This book follows a basic plan of presenting the material in approximately historical sequence. The first chapter traces the history of electronics (using a very broad brush!) in order to provide a background for understanding. Chapter 2 discusses power sources necessary for anything electronic to work.

Then in Chapter 3, we meet the passive circuit elements with which we will be working throughout the rest of our electronic careers. Once we know what the passive elements are, the next step is learning how to put them together into useful circuits.

The material presented in Chapter 4 is sometimes called "network theory" or "circuit theory" by jargoneers; it's the heart of everything that follows.

Chapter 5 introduces the active elements which make electronics capable of doing all that it does, and Chapter 6 puts them into circuits just as we put the passive elements of Chapter 4 into circuits with Chapter 5.

Finally, Chapter 7 rounds out the book with a handful of projects which were picked for their simplicity, general interest, and ability to backstop important points in the other chapters. By the time you've finished it, you should be well acquainted with the elements of practical electronics.

I hope you'll find the whole thing fun!

Jim Kyle

Editor's Note

Jim Kyle is perhaps one of the most prolific electronics writers of our time. He has authored hundreds of technical articles for the top-ranking amateur radio magazines, an impressive number of books on specialty subjects within the general field of electronics, plus the renowned—and acclaimed—four-book amateur radio "study guide" series published by 73 Magazine and TAB Books.

From Elektron to Einstein

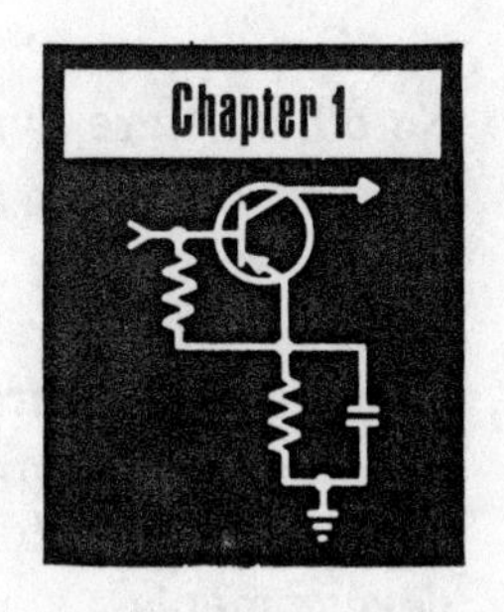

The Greeks had a word for it—and not much more. The Chinese put it to practical use—but they neither knew nor cared why it worked. Throughout most of recorded history it has been a magical subject enshrouded in mystery; only during the past century and a half has the mystery thinned enough to permit practical application of some of its elements.

Yet without this not-so-modern miracle, we could not have today's worldwide networks for instant communication of voices, pictures, and the written word. Home entertainment by television and recordings could not exist. The huge computers which control space flights—and the smaller ones which control our bank and charge accounts—probably would not even be dreamt of. The concentrations of commercial air transport which move people and products across the continent or an ocean in hours rather than weeks would be impossible. And this is only a partial list of the everyday things which would be affected.

That's the extent to which **electronics** affects our lives. Electronics is that branch of technology which deals with electricity and magnetism, plus their control and effects. As such, it is a part of the field of science known as **physics**. It's a relative newcomer, since the word **electronics** in its present meaning came into use only about 30 years ago.

Before that time, when the only familiar application of electronics was in radio, most folk thought of radio itself as a science.

In addition to "radio" there was "electricity"; and the people who worked with electricity were **electricians.**

It wasn't just popular speech that divided all of electronics into those two categories. Universities offered courses in electrical engineering, and the student had his choice of the "communications" curriculum or that dealing with "power." No others were available.

The foremost professional group dealing with this embryonic technology was known for many years as the **Institute of Radio Engineers**, and only within the last two decades has it managed a merger with the **Institute of Electrical Engineering** to form the modern **IEEE**.

Not until World War II brought into the open such diverse developments as radar, sonar, and magnetic recording—which clearly fit neither into radio nor power—did it become obvious that a more generic name for the technology was necessary.

Since all of the concerned developments worked on the basis of the electron theory, **electronics** was a natural as a name. It might surprise you to discover, though, that even today this term is not totally accepted. At least one leading encyclopedia has no data on electronics as such; its material on the subject is in an article titled **Electricity and Magnetism**. This technology, more than most, is still very much in the act of growing to maturity.

With a name less than 30 years old, most of its theories less than 150 years in age, and applications almost as new as tomorrow, it might seem almost ridiculous to hint at the "history" of electronics. However, the history exists, and it's long—3000 years' worth or more. One of the easiest ways to become familiar with the essential elements of electronics is to retrace that history in capsule form.

It began around 1100 B.C. in China, with the discovery that certain stones had the property of attracting other stones to them. Someone noticed that one of these stones, if floated (in a tiny model boat) on water so that it was free to move, would always point to the north, and the compass was born.

By 800 B.C., knowledge of this "lodestone" existed in Greece also. Its aid to navigators made possible early exploration of the seas, and aided the spread of civilization.

For Western peoples, one of the main sources of supply for this strange mineral was the district called Magnesia, near the Aegean Sea. In addition to the name "lodestone," it also began to be called "magnetic" material, and its strange property of attraction began to be called **magnetism**.

A little later, but still some 600 years before Christ, residents along the Baltic seacoasts had discovered a strange transparent golden stone which sometimes washed ashore after storms. This stone was soft enough to cut and carve into jewelry, yet hard enough to take a high polish.

But the property which made it so strange was that it would, when rubbed briskly, attract light objects such as straw and chaff, just as the lodestone attracted iron and other magnetic objects.

Because it would attract chaff, the Greeks gave it the name **elektron** (that which attracts chaff). Other peoples called it by other names. We know it today as **amber**.

From those times until the end of the 16th century, both lodestone and amber were generally thought to be magical substances and a substantial body of superstition grew about them. As William Gilbert, the founder of electronics as we know it, wrote in 1600, "Many philosophers cite the lodestone and also amber whenever, in explaining mysteries, their minds become obfuscated and reason can no farther go."

Gilbert also pointed out that most of the knowledge of his day concerning either amber or the lodestone had been passed on from earlier writers, rather than determined by experiment. It was Gilbert's extraction of truth from the mountain of legend that earned him the title of founder of the science.

He established, for instance, that the strange power of attraction found in amber also existed in many other substances, including most jewels, glass, sulphur, and hard resin. (Amber itself, we know now, is only fossilized resin from prehistoric pine trees.) He also laid to rest the myth that amber would attract chaff but not leaves, by proving that the attraction existed toward almost all light objects which we would now call insulators.

The attraction of the lodestone, on the other hand, which previous authorities had linked to that of amber, Gilbert assigned to a different cause entirely—thus establishing a separation between the study of electricity and that of magnetism that exists to this day in some circles, notwithstanding the fact that both are facets of the same set of circumstances.

Since historically the electric force of amber and the magnetic force of the lodestone were studied separately, in our retracing of history we must also divide out attentions. However, we must remember as we do so that no real progress in electronics was achieved until the relationship between the two was made apparent. But that is getting the cart before the horse...

After Gilbert's pioneering work was published in 1600, the next key event in our history was the discovery that all substances could be classed either as conductors or as insulators of electricity. This came to light in 1731, as Stephen Gray attempted to learn how the "electrick fluid" traveled. He found that when he suspended a thread by a dry silk line, he was able to transmit a charge along nearly 300 feet of the thread, but that when the silk was replaced by a brass wire, no trace of the charge could be detected. "By this," Gray reported, "we were now convinced that our former success depended upon the lines that supported the line of communication."

Although Gray's report was not published until 1731, the actual experiment was conducted on July 3, 1729. By this time, the modern scientific age was beginning to come into bloom, and much of the interest was in the area later to become known as electronics. One of the major questions puzzling the curious investigators at that point was how "electrick bodies" got to be that way.

One theory, put forward by the French researcher Charles Francois de Cisternay du Fay in 1734, held that there were "two distinct electricities, very different from one another." As du Fay explained it, "The first is that of glass, rock-crystal, precious stones, hair of animals, wool, and many other bodies; the second is that of amber, copal, gum-lack, silk, thread, paper, and a vast number of other substances."

The first of these he called "vitreous" and the other "resinous" electricity. "The characteristick of these," he wrote, "is that a body of the vitreous electricity, for example, repels all such as are of the same electricity; and on the contrary, attracts all those of the resinous electricity."

This theory, known as the "two-fluid" theory of electricity, enjoyed wide acceptance. One philosopher who disagreed, though, was a Philadelphia printer named Benjamin Franklin. He proposed a counter theory, known as the "one fluid" concept of electricity. According to Franklin, "electrical fire is a common element" equally distributed among all objects. If, however, one person while insulated from the common supply should collect the electrical fire from himself into a glass tube, and another person insulated both from the first and from the common supply should then draw fire from the tube, both the first and second persons would appear electrified to a third individual not insulated from ground.

The observation was one of fact; Franklin drew his theory to account for the observation. Rather than dividing all electricity into two classes as did the two-fluid theory, Franklin called the charges **positive** and **negative**.

As it turned out, he was wrong in only one particular. When science finally unraveled the hodgepodge of details governing the manner in which a charge could be accumulated, it turned out that the type of charge Franklin had named "positive" or "plus" actually results from a shortage of electrical particles, while the charge he called "negative" or "minus" is produced by a surplus! With this single exception, Franklin's theory turned out to be in strict accord with all later developments.

Throughout all this experimentation the only way the student could capture an electrical charge to investigate its action was to rub an insulating body briskly, and study the charge before it leaked away. But in 1746, experimenters at the University of Leyden in the Netherlands accidentally discovered the principle of the capacitor and its ability to store charges. (One account claims that the discovery was made by a layman who was amusing himself with the University apparatus.) The Leyden jar, which could hold an electric charge for amazing periods, was an indispensable part of all electrical experimenters' apparatus well into this century.

The Leyden jar was the last major invention of the golden era of electrostatics (the study of static electricity). Not long after its discovery, a source of dynamic electricity was uncovered and attention turned to work with it. But before we leave electrostatics, let's see why it works:

As early experimenters well knew, the simplest way to generate a static electric charge is to rub an insulator very briskly. This works best on a dry day. You may, for example,

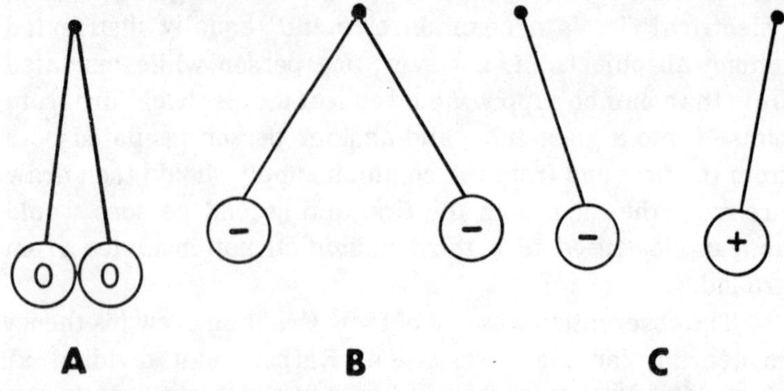

Fig. 1-1. One of the earliest electrical observations, which forms a key point to present-day circuit theory, was that objects with similar electrical charge repel each other, while those with opposite charge attract. Ping-Pong balls hung by threads from a support can demonstrate the effect. With no charge (A) two balls hang side by side. When both are given negative charge (B), they fly apart from each other. If two are suspended separately and given opposite charges (C), they attract—but if they touch, their charges will either neutralize each other and result in condition A, or one charge will overpower the other leaving both with the same charge (condition B).

have generated a startling charge yourself by sliding rapidly across the plastic seatcover of an automobile on a brisk winter day. If you hadn't immediately touched the ignition key or a door handle, perhaps you'd never have known about the phenomenon at all.

You can set up your own experiment to observe some of the effects of static electricity. Suspend a couple of Ping-Pong balls from the ceiling with 8 or 10 inches of thread in such a manner that the balls are about an inch apart when they are at rest. (See Fig. 1-1.) Watch what happens when you give one of the balls a negative charge and the other a positive charge. Then give both balls the same charge polarity. This demonstration can be impressive if you've never witnessed electrostatics and their effects.

Franklin described the result as a collection of the electrical fire into the body being rubbed. In actuality, the energy imparted by the rubbing carries electrons away from the insulator, leaving behind a shortage (Franklin's famous error). The reason it works only with an insulator is simply that if the material were conductive, the shortage would be replenished from ground (Franklin's "common supply"), and no charge could be detected.

Electrostatics is still an important branch of science, especially with regard to nuclear physics (and a lightning bolt is simply a vast electrostatic discharge). Today's electrostatic charges are often developed by a Van de Graaff generator (Fig. 1-2). This device uses a small charge injected onto a moving belt from point A, which is carried by the belt to a second point (B) inside a conductive sphere mounted on an insulating tube containing the belt. Because of the equalizing effect of electrical charge, every increment of negative charge imparted at B drives an equal amount away at C, producing a positive charge at C.

The positive charge from C is carried away by the belt, past point D to the grounded lower pulley. This charge, as it moves past point D, increases the charge on point A by induction, and so helps add more and more charge to the insulated sphere.

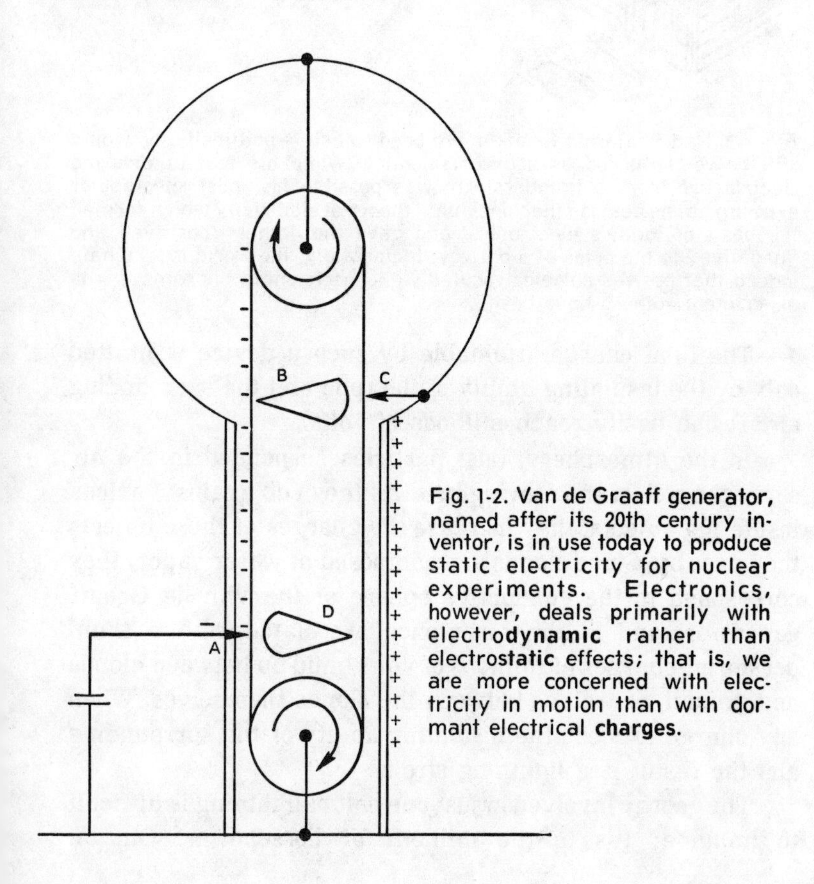

Fig. 1-2. Van de Graaff generator, named after its 20th century inventor, is in use today to produce static electricity for nuclear experiments. Electronics, however, deals primarily with electro**dynamic** rather than electro**static** effects; that is, we are more concerned with electricity in motion than with dormant electrical **charges**.

Fig. 1-3. Had Benjamin Franklin not been so active politically, he would still be remembered as a great scientist. While his feat of drawing electric fire from a thunderstorm was possibly his most spectacular experiment, he devised the "one fluid" theory of electricity which formed the basis of today's electronics, and gave the names "positive" and "negative" to the poles of a battery. Incidentally, the world is fortunate indeed that he was not electrocuted while performing his famous kite experiment; others have been.

The final charge attainable by such a device is limited only by the insulating ability of the tube and the surrounding air. It can easily reach millions of volts.

In the atmosphere, dust particles suspended in the air serve the role of the moving belt. As they rub against various insulating objects, they increase the charges of those objects they pass by. Since clouds are composed of water vapor, they correspond to the conductive sphere of the Van de Graaff generator, and in the turbulence associated with a thunderstorm, charges of millions of volts build up between clouds and ground, as well as between the clouds themselves. When any charge exceeds the insulating ability of the surrounding air, the result is a lightning stroke.

The energy involved in just one bolt of lightning is difficult to imagine; it's in the millions of horsepower. One of

Franklin's most noted experiments (Fig. 1-3) was his proof that lightning and electricity are one and the same; it's a mark of his unusual luck that he was not killed by that experiment despite his precautions. Eager experimenters are warned not to try to fly a kite in a thunderstorm; the forces are considerably beyond casual comprehension.

While Franklin was proving that electrical fire and lightning are identical, an Italian anatomist at the University of Bologna noticed that the legs of a frog he was dissecting would convulse whenever an electric spark passed from an electrostatic generator through the body. This led to Luigi Aloisio Galvani's interest in animal electricity—but the fundamental discovery which he overlooked had to wait nearly a decade for Alessandro Volta. As an anatomist, Galvani was convinced that the electric action of the frog's legs was due to the same kind of animal electricity possessed by the electric eel; he did not realize that it was due to the action of the external electric current.

Volta, a physicist, was studying the effects of simple mutual contact between metals of different sorts when he constructed an apparatus which produced an effect which, in Volta's words, "resembles Leyden jars...which act unceasingly" in a manner which "provides an unlimited charge." His invention was made known to the world in a letter to Great Britain's Royal Society dated March 20,1800. It was the dawn of a new day for electronics, because the "Voltaic pile" was the direct predecessor of today's dry-cell batteries; it made the study of "electricity in motion" possible for the first time. The golden age of electrostatics had ended; **electrodynamics** was born.

Two hundred years previously, Gilbert had separated the study of electric and magnetic effects. The separation was well justified, because static electricity and static magnetism are independent of each other. Nevertheless, the similarities between electric fields and magnetic fields were enough to assure that students interested in one would also be curious about the other.

The period from 1600 to 1800 was primarily one of discovery concerning electric effects. Knowledge concerning magnetism remained at about the same level during this era.

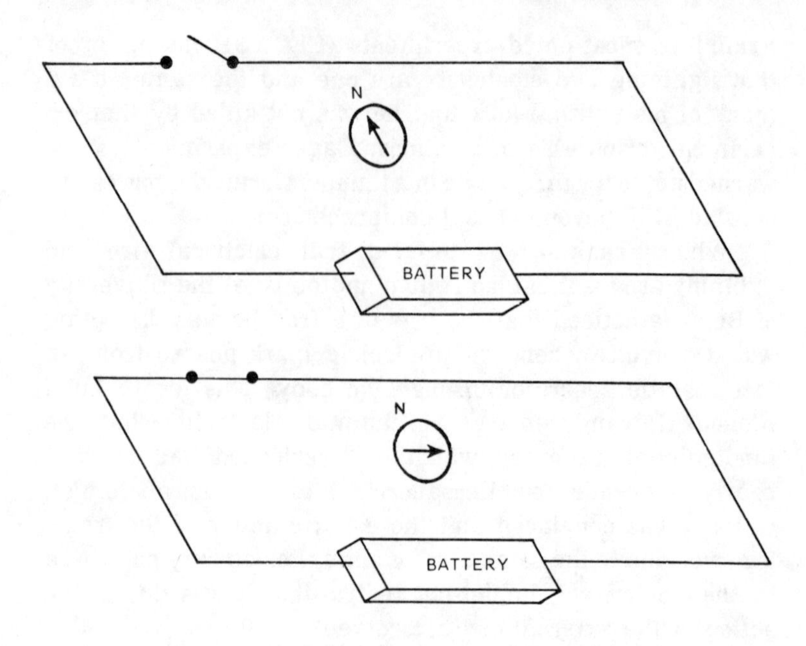

Fig. 1-4. Oersted's great discovery was that magnetism and electricity were linked to each other. He found that a compass needle which normally pointed north (top) was deflected from its normal position when current was allowed to flow through a nearby conductor (bottom). No physical connection between electric circuit and compass existed; yet, the compass was affected. The only possible conclusion to be drawn, was that the sudden flow of current somehow caused a magnetic field to "appear." Later workers cast doubt on a "cause and effect" relationship between electric and magnetic fields, but they are always together whenever current is in motion.

The one major exception to this was the work of the French engineer-physicist Charles-Augustin Coulomb. Coulomb first won fame in the field of physics with his discovery of the torsion-balance principle, which was the forerunner of today's torsion-bar suspension for automobiles. Coulomb's torsion balance made it possible to measure microscopic forces, and he turned his attention to measurements of the forces involved in attraction and repulsion by both electric and magnetic fields. Results of these studies were his formulation of the law of electric force; "The repulsive force between two small spheres charged with the same sort of electricity is in the inverse ratio of the squares of the distances between the centers of the two spheres," and of the law of magnetic force, which showed the same to be true of magnetic repulsion, with addition of a modifying influence: the strength of the magnets involved.

Coulomb's work was done in the last part of the 16th century, just before Volta's invention of the chemical battery. It pointed the way directly toward a later rejoining of the study of electricity and of magnetism. In 1820, the Danish investigator Hans Christian Oersted demonstrated that the magnetic needle of a compass could be moved from its position by making and breaking an electric circuit (Fig. 1-4) from a galvanic battery. This was subsequently explained by French physicists Biot and Savart as a magnetic field surrounding an electric current.

Oersted's experiments led Dominique-Francois Jean Arago to try to induce magnetism in iron by means of electric current, and he did so successfully. Arago then went on to discover that no battery was necessary in order to interfere with a magnetic field; he found that a moving conductor all by itself would interact with a magnetic field in a manner which posed a deep mystery to science for several years; more importantly, the strange phenomenon inspired additional experimentation.

The decade 1820-1830, which began with Oersted's relinking of magnetic and electric phenomena, set the stage for an age of discovery and application. Andre Marie Ampere, a French mathematician, studied Oersted's results and developed a theory to account for them. In the course of his studies, he gave to many phenomena the names which we use today—in particular, he established the terms **electrostatic** and **electrodynamic** to differentiate the older and newer forms of electricity, and made a special point of showing that the magnetic effects were due to the motion of the electric charge, while the electric attractions and repulsions which had been studied for so long were due to unequal distribution of electricity at rest in the bodies in which they were observed.

Then in 1826, a German mathematician and schoolteacher by the name of Georg Simon Ohm published a study of resistance in electrical conductors, which concluded in the famous law which bears his name and links voltage, current, and resistance into a single interrelated phenomenon.

Meanwhile, in England, a young bookbinder had become interested in science and had gained a position as assistant to the noted physicist Sir Humphrey Davy. In 1825 the assistant was made director of the laboratory of the Royal Institution, and on November 24, 1831, Michael Faraday read to the Royal Society the first of his long series of **Experimental Researches in Electricity**. They continued through 1854, and in their course, many of the major developments of electronics occurred.

One of the first—and, many believe, the most important—of Faraday's discoveries was the action he called **electromagnetic induction**, which accounted for Arago's moving-conductor mystery, and went beyond this far enough to make possible the discovery of the dynamo.

The first step of this discovery was the invention of what we now know as the transformer—two coils of wire, insulated each from the other, but physically close together. Faraday observed that when direct current flowed through one of the coils, a brief flicker of electricity appeared in the other. When the circuit was connected to the first coil, the flicker was always in the same direction, and when the circuit was disconnected, the flicker went in the other direction. This led him to the belief that the battery current through one wire did indeed induce a current in the other wire. Actually, the current's change from zero to some established value caused a magnetic field to be established, and the change in this field from zero to its final value in turn induced the current in the second wire, but this refinement of the explanation had to await the next step of his experimentation.

That next step was the inclusion of a soft iron core in his primitive transformer. The flicker was increased "to a degree far beyond what has been described when with a battery of tenfold power, helices without iron were used," Faraday reported. Since Arago had already shown that passing current through a coil of wire around an iron core would induce magnetism in the core, this observation by Faraday established a connection between magnetism and the induction of electricity. His next step was to try an ordinary magnet instead of an electromagnet.

"By breaking the magnetic contact" in his apparatus, Faraday was able to reproduce the electric induction at will.

Now having proved that electricity could be produced from action of a natural magnet, without chemicals, the great scientist made one of his major contributions. He demonstrated that the effect was due not to "some peculiar effect taking place during the formation of the magnet," but only to its movement, by suddenly thrusting a magnet into the coil. The meter flickered.

Thus he proved that an electric current is excited in a conductor when a magnetic field moves in its vicinity, and since motion is relative, the same is true of a moving conductor near a magnet.

This discovery led directly to the dynamo, the first electric generator, which converts mechanical motion into electrical energy by using the motion to carry a magnet past a coil (or vice versa). The automobile generator, as well as virtually all commercial electric power plants, still make use of this principle.

Faraday continued his study of electricity throughout his life, and went on to discover the principle of self-induction, which has formed the basis of most automotive ignition systems. This same principle was independently discovered by the American experimenter Joseph Henry, who later became the first secretary of the Smithsonian Institution.

Henry's major contribution to electronics, however, was his discovery of **resonance**, and his comprehensive explanation of its action. This phenomenon is a transfer of energy from electrostatic storage (such as a Leyden jar) into a magnetic storage (such as the effect of self-induction), back into electrostatic storage, and so forth until all the stored energy has been dissipated. The oscillation of energy from electric to magnetic storage and back again made it possible to transfer energy, without wires, over a distance of 30 feet and through two floors and ceilings each 14 inches thick.

By this time the discoveries and observations made by Faraday, Henry, and their contemporaries had thrown classical physics into turmoil. Based as it was upon Newton's laws of mechanics, physics could not at that time account for actions between physically separated objects. Yet induction obviously was an effect due to "action at a distance," and just as surely did exist.

In 1865 the Scottish physicist and astronomer James Clerk Maxwell published a "dynamical theory of the electromagnetic field," in which he put forth a possible explanation of the new observations. Maxwell's theory, which he expressed finally in terms of 20 separate equations in 20 variables, provided an explanation which has proved adequate from that time forward, even though several of the points on which it was founded have since been discarded!

The famed equations, prerequisites of knowledge for an electronics engineer, are not simple. In the century since they were first published, the original twenty have been boiled down to four—but in the process of doing so, the resulting four became entangled with integral calculus, differential equations, and vector analysis.

Taken apart from the equations, however, the theory is elegantly simple and is not difficult to understand. We must keep in mind, however, that it cannot provide answers to a few critical questions. The theory begins with a single assumption—that the space around objects which exhibit electric or magnetic effects contains "matter in motion" (to quote Maxwell) "by which the observed electromagnetic phenomena are produced."

From this starting point, Maxwell defined "the electromagnetic field" as "that part of space which contains and surrounds bodies in electric or magnetic conditions," carefully pointing out that even though this might be as complete a vacuum as could be obtained, there would still be "enough of matter left to receive and transmit the undulations of light and heat." To do so, he drew upon his background in astronomy to point out that we receive both light and heat from the sun, **through the vacuum of empty space**!

Because the transmission of light and heat remained essentially unaffected when transparent objects took the place of the vacuum, Maxwell concluded that the undulations must be taking place in some form of matter not directly observable. His term for it was "an aethereal substance." Later workers shortened this term to "the aether" or "the ether"; later studies cast grave doubts about its existence, which Maxwell's theory could not answer, being based as it was upon existence of this ethereal substance. The theory, however, still holds—whether or not the ether has any real existence; and it can be comprehended most easily by assuming (at least temporarily) that there is such a substance.

Since a detectable time delay exists between departure of energy from a source and its arrival at any other point, Maxwell concluded that "the energy communicated must have formerly existed in the moving medium." Borrowing from established knowledge in the field of mechanics, he theorized that while the energy was in the unknown medium, it must have been "half in the form of motion of the medium, and half in the form of elastic resilience." Figure 1-5 shows the idea.

This, in turn, led to the conclusion that the parts of the unknown medium must be connected in such a manner that "the motion of one part depends somehow upon the motion of all the rest," and at the same time "the connections between the parts must be capable of elastic yielding" of some type, else transmission would be instantaneous rather than requiring a short but detectable time delay.

Having drawn this conclusion, Maxwell reworded it into a description of the significant properties of his postulated medium. Its major property was the capability of "receiving and storing up two kinds of energy, namely, the **actual** energy depending on the motion of its parts, and **potential** energy, consisting of the work which the medium will do in recovering from displacement in virtue of its elasticity."

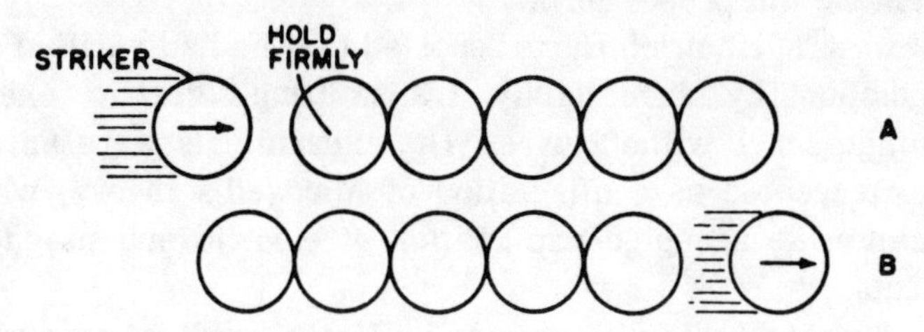

Fig. 1-5. Row of coins offers a model of Maxwell's theory of electromagnetism. If row of coins is placed end to end and one end struck by another moving coin, as at top, motion of the striker coin (row A) is transmitted through entire row with little loss and imparted to the extreme-right coin as shown at bottom (row B). Similarly, Maxwell believed that electromagnetic energy was transmitted from point to point at high speed by series of similar compression-expansion cycles of adjacent points. The popular "swinging ball" toys sold as executive "pacifiers" also serve to demonstrate this interesting transfer-of-energy principle.

These two kinds of energy correspond exactly to what we now know as **kinetic** and **potential** energies of moving objects—and also, which was the point of the whole theory, to the **electrodynamic** and the **electrostatic** effects distinguished by Ampere.

Maxwell's equations describe these properties in mathematically precise detail. In the process, the concepts of electric fields, magnetic fields, and **waves** are made precise for the first time. We'll get on to the details of the theory in the next chapter—right now, let's see what happened between Maxwell's 1865 report and the present day.

The first major effect of Maxwell's theory was to give science a tool for **predicting** events which had not yet been observed. This is the primary test of any scientific theory, and Maxwell's ideas were put to the test soon after they were published. The equations indicated that heat, light, and electromagnetic energy were all different manifestations of the same basic kinds of events. Since both heat and light could be sent through empty space, without wires, and the theory held that electromagnetic energy was similar, then electromagnetic energy should also be able to travel without wires. All that was required was apparatus for putting the energy into the proper form for transmission, and a means of detecting its reception. The theory indicated that the "oscillatory discharge" discovered previously by Henry might be the proper form.

In 1888, Heinrich Hertz reported that he had verified this prediction by successfully transmitting electric energy through space without wires. His experiments were universally accepted as confirmation of Maxwell's theory, which became the accepted explanation of electric and magnetic effects.

A young Italian, impressed by Hertz' work, attempted to duplicate it. When he had done so, he turned his attention to making practical use of the apparatus, and invented the antenna or aerial. This made it possible to send messages without wires. Marchese Guglielmo Marconi then astounded the world by sending a wireless message across the Atlantic ocean; in doing so, he opened the door for commercial radio and its related industries.

Meanwhile, back in the laboratories, Sir William Crookes discovered an electrical discharge tube which became known as the cathode ray tube. This discovery led many workers into study of the electrical nature of matter in general. They quickly found that the cathode rays behaved just like charges of electricity having negative polarity. One discovery led to another, until publication by Sir J. J. Thomson of the "electron" theory in 1897.

This theory (which is today accepted as fact by so many persons that our reference to it as "theory" may come as a shock) explains not only electricity, but all physical matter, in terms of **atoms** (Fig. 1-6) which are, in turn, composed of **atomic particles**. The theory recognizes three types of such atomic particles: the **electron**, the **proton**, and the **neutron**.

According to the theory, the electron is a tiny particle of energy or matter, with a "negative" electric charge. In fact, it is usually called "the unit of electric charge" and all other

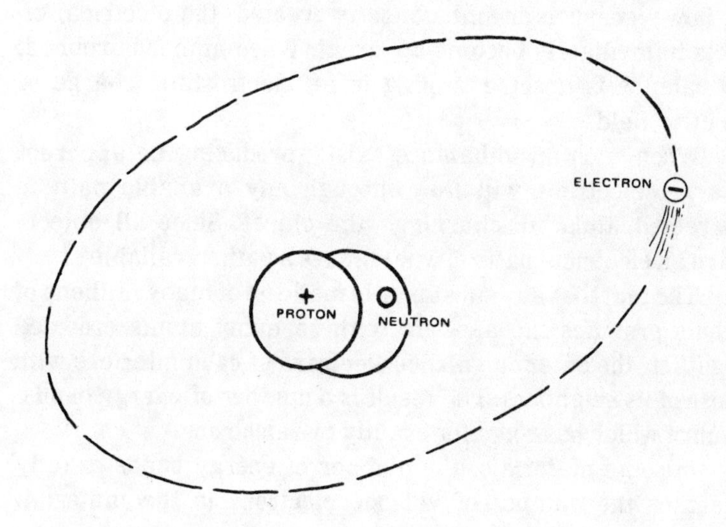

Fig. 1-6. According to present atomic theories, all matter is composed of molecules, which are in turn made up of atoms. An atom is the smallest particle of a chemical element capable of retaining the properties of that element; but atoms in their turn are composed of atomic particles called neutrons, protons, and electrons. The neutrons and protons make up the "nucleus" of the atom, while the electrons are in "orbital shells" at a relatively great distance from the nucleus. Electrical effects are due entirely to action of electrons. Shown here is a not-to-scale sketch of a helium atom, the simplest one which contains all three types of particles; the hydrogen atom is similar but has no neutron.

measurements of electric charge are made in terms of the number of electrons required to equal the charge being measured.

The neutron, much heavier than the electron, is so called because it is electrically neutral. The proton, with the same mass as the neutron, has a **positive** charge. (Some versions of the theory consider only two particles, disposing of the neutron as a bonded pair consisting of a proton and an electron in which opposite charges cancel.)

Since all matter is composed of atoms in some kind of structure, it follows that all matter must contain electrons and protons. If these were always perfectly balanced, the charges would always cancel and we could not observe any electric effects. What we know as an electric charge is produced by an imbalance between the number of electrons and the number of protons in any specific location at a specific time.

In most objects, the imbalance does not exist. That is why most objects ordinarily exhibit no electrical characteristics. If, however, such an imbalance is created, the electrical effects immediately become apparent. When amber is rubbed, its balance is upset, resulting in an electrostatic charge or electric field.

When such an imbalance exists, producing an apparent charge, electrons will flow through any available path to correct it, thus "discharging" the object. Since all objects have an electrical nature, what makes a path "available"?

The fact that any substance is made up of many millions of atoms provides the answer. With so many atoms crowded together, the outer or **valence** electrons of each interfere with those of its neighbors. The result is a number of **energy bands,** each of which has room for exactly two electrons.

In some materials, the number of energy bands exactly matches the number of valence electrons in the material. Electrons may drift from one atom to another only by replacing each other; no net movement can be detected as a result.

In other materials, the atoms are arranged so that there is either no gap between a full valence band and the next higher one, or only one electron per band. In this situation, electrons are free to wander from atom to atom as soon as any electric force is applied.

We call the first kind of substance an **insulator** and the second kind a **conductor**. Note that in either, individual electrons are free to move, but the results differ drastically.

Still a third kind of material exists. It's called a **semiconductor** because it behaves somewhat like a conductor and somewhat like an insulator.

In an insulator, unoccupied energy bands do exist, but are too far removed from the valence bands to have any effect at normal energy levels. As temperature rises, the gaps between bands become narrower and the insulator begins to behave more like a conductor.

In a semiconductor, the arrangement is like that of an insulator but the gaps are already closer together, so that some conduction occurs at ordinary temperatures.

All three kinds of materials offer some resistance to electric current, dissipating part of the energy by changing it into heat. Insulators have very high resistance; most pure semiconductors have moderate to high resistance; conductors have relatively low resistance.

One of the quick distinctions between conductors and semiconductors is what happens to their resistance when heat is applied. That of a conductor rises; that of a semiconductor drops. (We'll delve into this more deeply in Chapter 5.)

Acceptance of the electron theory as the basis of physics is what led to the name "electronics" being applied to our art. Thomson, in turn, chose the ancient Greek word as his name for the unit of electric charge, to bring the wheel of history full circle.

But that doesn't quite bring us all the way up to the present day. As the electron theory became firmly established, a number of troublesome questions arose. Until then, physicists thought of electricity as a fluid which could flow through all space. Maxwell's theory depended upon an unknown substance (ether) which had the physical properties of a fluid. This led to such concepts as current, waves, flow, pressure, and the like.

Quite naturally, experimenters turned their attention to this mysterious substance which supported electromagnetic waves. The mechanical properties of the "ether" were, at first, a mystery. Then H. A. Lorentz discovered that all the phenomena of electromagnetism then known could be explained by only two assumptions: that the ether is firmly fixed in space, unable to move at all, and that electricity is firmly lodged in the mobile atomic particles. This statement was paraphrased by Albert Einstein in 1934 as follows: "Physical space and the ether are only different terms for the same thing; fields are physical states of space."

Lorentz' discovery made little headway toward acceptance until Einstein published in 1905 his special theory of relativity, which provided a mathematical basis for equating space and the ether. The famous "ether-wind" experiment made in California by Michelson and Morley, which indicated that the earth must be at rest with respect to the ether, had already provided an observational basis. By merging the two distinct ideas of absolute time and absolute space into a four-dimensional space-time continuum, Einstein extended the concept of physical space to one which could include Lorentz.

Meanwhile, quantum mechanics and nuclear physics were being invented by Planck, Bohr, and others. The resulting "new physics" of the first part of this century posed new problems. No longer was it possible to think of an electron as a neat little packet of electric charge, or of a light beam as a wave of definite frequency. In some cases, an electron behaves like a wave rather than a particle. In others, light can be shown to consist of particles rather than continuous waves—but these particles may be in several places at the same time!

These problems remain as mysteries in today's science, which is usually disposed of by a tacit suggestion not to worry about it. Scientists know that neither the wave notion nor the particle concept is actually correct. Each, however, is highly useful in the proper circumstances, and so both remain in use although more or less universally acknowledged to be not quite "right." For practical purposes, we can only follow this same course.

Now, with this short history of physics behind us and a general knowledge of how electronics came to be, we can move into more detail. The names of many of the experimenters we have mentioned in this chapter will become familiar to you as we proceed, if they are not already, since many—indeed, most—of the pioneers of this science have been remembered in the naming of units of measurement.

The volt, for instance, which is the unit of electrical pressure, is named for Volta. Ampere's name went to the unit of electric current. Charge is measured in coulombs, and resistance in ohms. The unit of inductance is named for Henry. That of capacitance is derived from the name of Faraday, but the unit (the farad) turned out to be impractically large; the microfarad (millionth of a farad) is the practical unit. The unit of frequency, once expressed as **cycles per second**, is now referred to as **hertz**. There are gilberts and maxwells and watts and oersteds...and more. We'll meet at least a few as we proceed with our study of the elements of practical electronics.

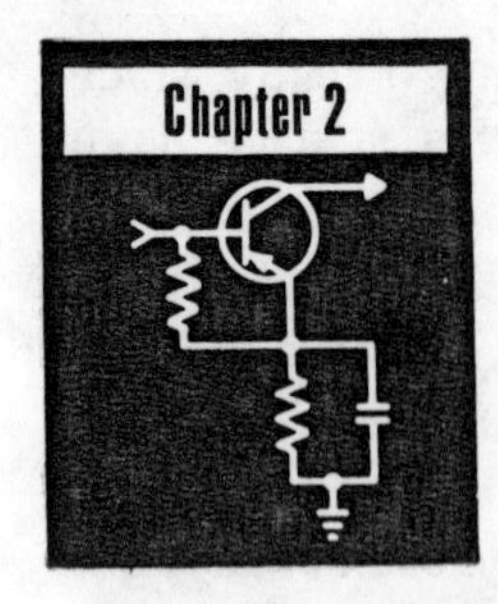

Energy Sources

In the previous chapter, as we traced the history of electronics from ancient times to the present day, it became apparent that the development of electronics really got nowhere until a reliable energy source was invented—after which things happened fast.

Actually, of course, energy has only one source (and science does not yet know what that is), since the product of energy and matter in the universe is constant and can neither be increased or diminished, although it is possible to transform energy to matter and vice versa as well as the more familiar transformations of energy from one form to another.

Since modern physics tends to be oriented in the direction of electronics, it should not be too surprising that the electrical form of energy is now considered to be an almost pure measurement of it. Atom smashers, for example, are rated in terms of the number of millions (or billions!) of electron-volts they can develop. The older forms of energy include mechanical motion, gravitational force (the potential energy locked in the pond above a waterwheel), and heat.

The major "sources" of electrical energy (we might as well follow common usage and call them sources, even though we know they actually are only agencies of transformation) are chemical cells and mechanical devices based on the dynamo principle. Less important sources today, though the situation may change in the near future, include fuel cells and atomic power packs.

Let's examine both the chemical sources and the dynamo-based sources in some detail, and then take a look at the less important means of obtaining electricity to power our electronics projects.

CHEMICAL SOURCES OF ELECTRICITY

The development of electricity and electronics as an applied science dates almost directly from Volta's invention of the voltaic pile—the first chemical cell. Until that time, the only source of electrical energy available to the experimenter was static electricity, and all the phenomena of current in motion were hidden from us.

The original cell described by Volta in 1800 (Fig. 2-1) was "only an assemblage of a number of good conductors of different sorts arranged in a certain way." He recommended 30 to 60 pieces of copper or silver, each in contact with a piece of tin or zinc, and each pair of metals separated by a blotter or piece of leather soaked in saltwater or lye. When these three sets of items were arranged so that they alternate by triplets, the result is a rather astonishing battery.

This original voltaic pile can be duplicated with coins. Alternating silver coins such as pre-1964 quarters with copper pennies and saltwater separator pads provides a battery with enough power to drive a simple transistor oscillator or light a low-current flashlight bulb.

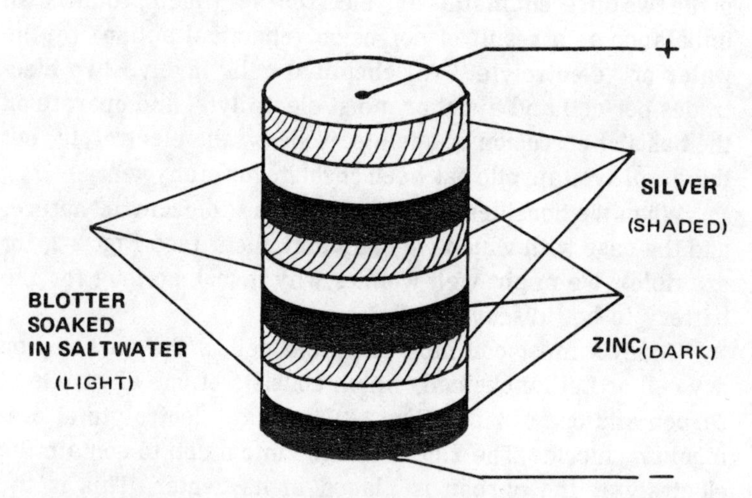

Fig. 2-1. The first source of steady current available to electrical experimenters was the "voltaic cell" shown here. It consisted of a stack of alternating zinc, blotter, and silver discs (blotters were saturated with saltwater). You can make one using silver coins and the more recent zinc-copper types—or even substituting copper pennies for the zinc. A silver-copper stack will develop about one-half volt per cell. The stack shown here contains three cells (each cell is a complete set of zinc, blotter, and silver), and thus provides a voltage output of about 1.5 volts—about the same as a flashlight battery.

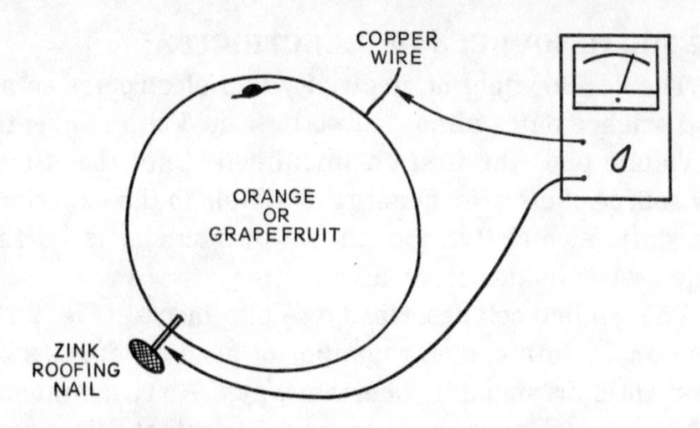

Fig. 2-2. One of the easiest-to-make batteries is the "citrus cell"—a lemon, lime, grapefruit, orange, or tangerine with a zinc-coated roofing nail driven into one side and a length of solid copper wire pushed into the other. These can be connected in series or parallel until the voltage and current are built up to useful values. As an experiment, try this with one fruit. Set your voltmeter to its most sensitive dc voltage position and watch the pointer jump when you touch the test leads to the fruit's terminals.

Most of the early discoveries in electronics mentioned in the previous chapter were made with the aid of such apparatus. It works because of the difference in electron density in the two different metals or "electrodes," which produces an imbalance as a result of corrosion (chemical action) by the water or "electrolyte." All chemical cells involve two electrodes per cell and a wet or moist electrolyte, and operate on the basis of corrosion of one electrode by the electrolyte, but the simple voltaic pile has been obsolete for many years.

When we consider the cells that tend to occur in nature, and the ease with which we can make them (see Fig. 2-2, for example), we might well wonder why it took so long for the battery to be "discovered."

Today's most common chemical cell is the carbon-zinc dry cell or LeClanche cell, which consists of one electrode of carbon and one of zinc (Fig. 2-3), with an electrolyte of ammonium chloride. The zinc is formed into a cup to contain the electrolyte; the carbon is placed in its center. This is the common flashlight battery.

Actually, the carbon electrode is not really carbon at all; it is composed of a mixture of manganese dioxide, carbon, and solid ammonium chloride. About half the energy output of the cell comes from decomposition of the manganese dioxide; the carbon is merely a conductor and does not take part in the chemical reaction. As the manganese dioxide is used up, cell output decreases. We say it "runs down" with use.

Other types of dry cell include the alkaline-manganese cell, the silver-oxide-and-zinc unit, silver-oxide-and-cadmium cells, and mercury batteries.

The chemical reaction in the alkaline-manganese cell is the same as in the carbon-zinc cell, but takes place in an alkaline mixture rather than the acid mixture used in the carbon-zinc unit. The result is a cell which provides up to twice the current output, with much longer shelf life.

The silver oxide units are based on Volta's original silver conductor. When used with zinc, they provide very high output for short times. They are most often used in military applications where long life is unnecessary (torpedo drive, etc.). With cadmium, silver oxide provides a very-long-life cell.

The mercury cell, unlike the alkaline and silver cells, features a relatively constant voltage rather than high power output. Its basic chemical reaction is between an

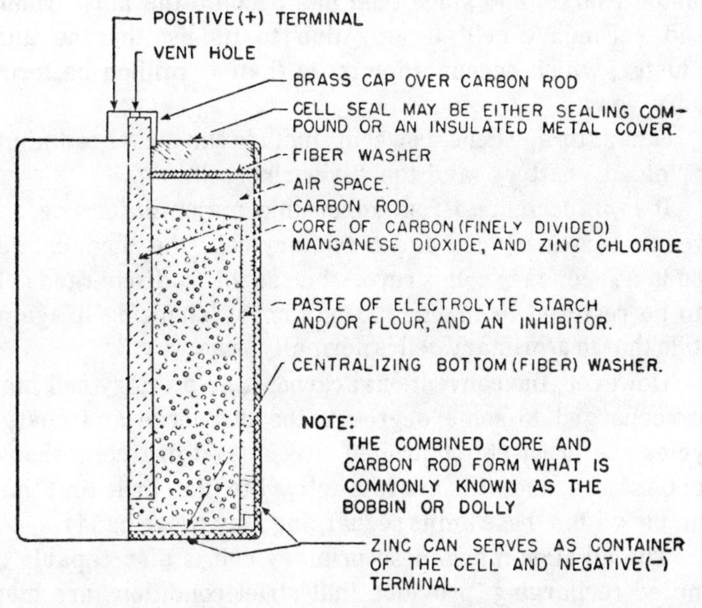

Fig. 2-3. Cutaway view of zinc-carbon cell (flashlight battery) shows carbon-rod inner electrode, surrounded by a mix of fine carbon, manganese dioxide, and zinc chloride, all packed into zinc can which is then encased in a steel outer jacket for mechanical strength and protection. Zinc can is corroded away during life of cell; without steel case, electrolyte would escape when cell runs down and destroy equipment using it.

amalgamated zinc electrode and a mercuric oxide electrode, with a potassium hydroxide solution serving as the electrolyte. Its electric energy results from controlled corrosion of the zinc by oxygen released from the mercuric oxide. A byproduct of this corrosion is the availability of free electrons at a potential of about 1.3 volts.

All these chemical voltage producers we have discussed so far are what are known as **primary** cells. That is, they produce power whenever a load is connected, without any requirement for being charged first. This is possible because they release electrons as a byproduct of controlled corrosion. A **primary** cell always has a "consumable" electrode that degrades gradually during use and so is not capable of being restored to its fully charged state.

Storage cells, on the other hand, store electrical energy in the form of chemical action, and release it later. The everyday automobile battery is a typical example of the lead-acid storage cell. The lead-acid cell was first invented in 1860 by Gaston Plante, and since then has become the most widely used **secondary** cell—largely due to its use by the auto industry, which accounts for more than 41 million batteries every year.

Other storage cells, however, include the nickel-cadmium or "nicad" battery, and the silver-zinc cell.

It is important to remember this major difference between a secondary cell and a primary cell: the chemical action in a secondary cell is reversible, so that a discharged cell can be restored to like-new condition by charging it again, while that in a **primary** cell is normally "one-way."

However, the conventional carbon-zinc primary cell may be recharged to some degree if the discharge and charge cycles are precisely controlled. Recharged dry cells should not be stored, and will have less active life than their first time out, but within these limits recharging may prove handy.

The alkaline-manganese primary cell is also capable of limited recharging, provided that strict conditions are met. The silver-oxide cells, as we have seen, can be used interchangeably as either primary or secondary cells, but in either case are costly.

In the lead-acid secondary cell (Fig. 2-4), both electrodes consist of lead compounds, and the electrolyte is usually sulfuric acid. The negative plates are a special spongy lead, and the positive plates consist of lead dioxide. The lead of the negative plates reacts with the acid to form lead sulfate, and in the process electrons are freed. At the same time, the lead dioxide of the positive plates accepts electrons while reacting with the acid to form lead sulfate also. Thus two different reactions, with opposite polarity, are going on at the same time, and the result is a potential of about 2.2 volts.

When recharging, the current flow through the acid (which retains the lead sulfate in solution) tends to decompose the lead sulfate. Lead atoms are deposited on the negative electrode, and lead dioxide is formed at the positive plate. This is an exact reversal of the action taking place upon discharge, so that the electrical energy is stored.

The familiar car-battery outline is not the only one possible for the lead-acid cell. Its high power storage capacity

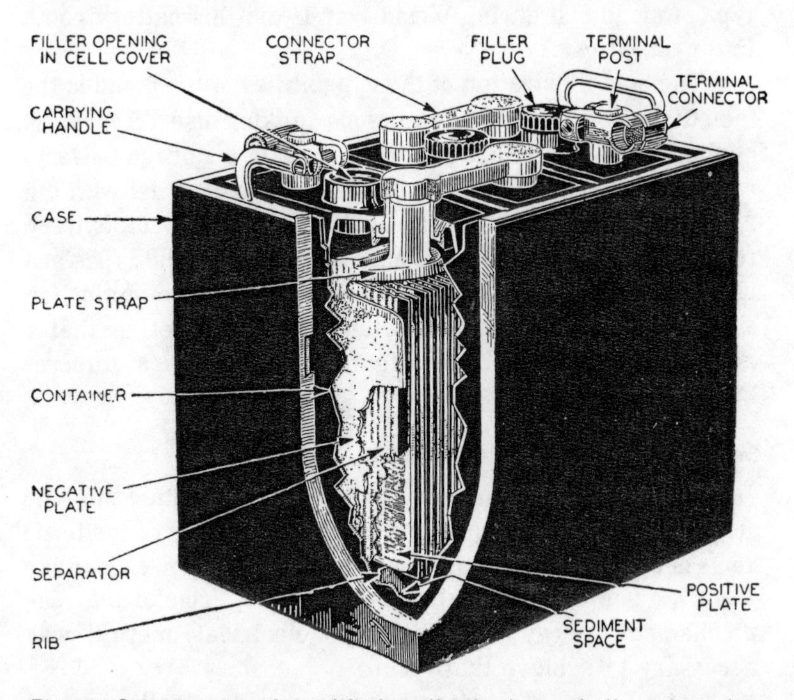

Fig. 2-4. Cutaway view of simplified 6-volt auto storage battery shows two sets of plates interleaved in each cell. Three such cells are connected together to form a **battery** of cells. Each set of plates converts electrical energy to stored chemical energy during charging process, and releases it back to electrical form during discharge cycle. Usual electrolyte for auto battery is sulfuric acid. Plates are made of lead. Electrolyte dissipates during use and must be added through filter openings.

(an auto battery will supply 3600 watts for 10 to 15 minutes, while starting the motor) makes it a natural for cordless tools such as grass shears and power drills.

In these applications, the lead-acid battery is used in a "sealed" form, with its acid electrolyte immobilized. Power storage ability is not significantly reduced, staying at about 15 watt-hours per pound of battery weight.

The major competitor to the lead-acid cell in storage applications is the nickel-cadmium or **nicad** battery, which sparked the boom in cordless appliances. The nicad uses one electrode of nickel and one of cadmium, in an electrolyte of potassium hydroxide. The first such cell was developed in 1899 in Sweden, and the original design is still in use as a heavy-duty workhorse. However, it was a heavy battery which found its major applications in emergency standby power applications. Not until the Germans developed the sintered-plate type of nicad cell during World War II did this battery come into general use.

The best illustration of the capabilities of the nicad is the fact that possibly half the persons making use of cordless appliances do not realize they even contain a storage battery.

The nicad cell gives 1.2 volts per cell, in contrast with the 2.2 volts provided by each lead-acid cell. Thus, a 6-volt battery requires five nicad cells, but only three if lead-acid types are used. Either type of cell can provide high power output for short periods of time. Manufacturers of nicads tell us that a nicad half the size of a flashlight cell, rated at 0.8 ampere-hour, can deliver 10 amperes for a few seconds.

MECHANICAL ENERGY SOURCES

While the direct current provided by the voltaic pile and its successors was adequate for discovery of most of the basic facts of electronics, it could not provide enough power to make electric power practical. The dynamo, converting the mechanical energy of motion into electrical energy, was necessary to achieve that result.

Not all of the difference between the chemically derived electric power and that obtained from mechanical energy is obvious. One of the most important differences has to do with the polarity of the energy flow, which has all manner of side effects.

The power derived from chemical action, whether in a primary cell or from a storage cell, is a continuous action. Its direction never changes while the power is being used. The positive terminal of the battery is the positive terminal all of the time, not just now and then.

The power derived from mechanical motion, on the other hand, can and often does reverse its polarity at regular intervals. This apparently minute difference makes it possible for such power to be carried efficiently for hundreds of miles across the country, from Niagara Falls to New York City, or from Hoover Dam (near Las Vegas) to Los Angeles. Since it has such a major effect, it's worthwhile to take a look in more detail at just how such a difference comes to be. The starting point is with a flow of electric current.

We've already met several methods of causing an electric current to flow. We can charge an insulator with static electricity, or put together any of several chemical reactions.

No matter how we get the current to flow, what we are dealing with is essentially a batch of electrons in motion. Since only the "free" electrons in any element are able to move readily, current can flow only through electrical conductors. Actually, the electrons themselves do not move at all rapidly. The average man out for an early morning jog moves with more velocity than the electrons in a current-carrying wire. What happens is that each electron as it moves hits others, and the force of impact ripples along the conductor far more rapidly than the electrons themselves travel. The effect, then, is that electromagnetic energy travels at the speed of light.

This is as good a time as any, incidentally, to point out again that Ben Franklin, when he assigned the signs "positive" and "negative" to the two polarities of electric charge, missed. The actual electron charge is the kind Ben called minus. The net result of this is that in the electrical circuit, **electrons in motion which constitute the current are "flowing" from negative to positive all around the path.**

Physics, however, in the years before Franklin's mistake became obvious, developed a considerable mass of theory and teaching which held that current flow is from positive to negative. This contradiction in terms has brought many students of electronics to grief. All we can do about it is to remember that current flow is generally "thought of" as being

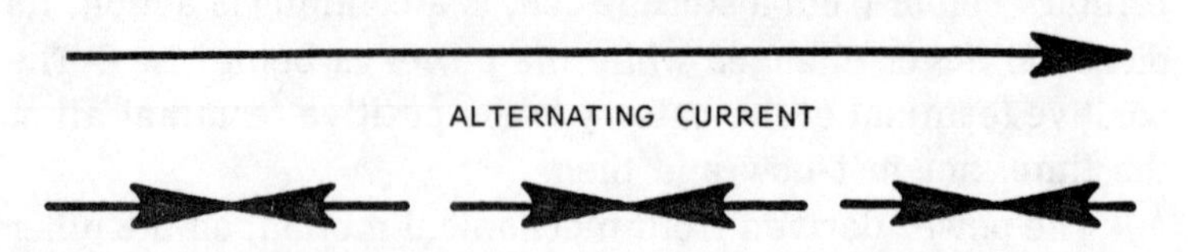

Fig. 2-5. Difference between direct current, obtained from chemical sources, and alternating current, obtained from mechanical motion, is that direct current always flows the same way while alternating current reverses its direction of flow at regular intervals. Alternating current (ac) can be converted to direct current (dc) by either mechanical or electronic means, so that mechanical generators can appear to produce dc, but it's always ac originally.

from positive to negative, while actual **electron flow** is from negative to positive. Confusing, yes, but it's something we must learn to live with until the books get changed and electrical symbols are redrawn to more accurately depict the directions in which electron impacts occur.

The flow of current (or electrons) is measured in units named for Ampere. The ampere is defined directly in terms of the amount of charge which passes a given point within a specified time. One ampere is the current which will carry one coulomb of charge past the measuring point in one second. The coulomb, in turn, is defined as 6.28×10^{18} electrons worth of charge, but so far no one has stopped to count the electrons. For one thing, the ampere is also defined in terms of its magnetic effect, and that's the definition used to actually measure the "standard" ampere.

The key thing to remember about current is that **time** is an essential element in its definition. It's easy to fall into the trap of thinking of current as a measure of the amount of charge; actually, though, it's a measure of the amount of charge **per unit of time**. It's like the difference between distance, say in miles, and speed, in miles per hour. Unless the charge is moving, current cannot exist. And if electric charge is moving, current flows.

It's possible, for instance, to move a quantity of charge a microscopic distance in one direction, then reverse its motion as shown in Fig. 2-5 so that it retraces its path, and repeat this action continually so that the actual average movement of the charge is zero—but a current flows, just the same, because charge is in motion.

And that's the difference between the chemical-derived current which always moves the same way, and that obtained from mechanical motion, which can readily be made to reverse or "alternate" its direction at regular intervals.

Regular household power, in fact, is just this kind of alternating current, while the current from a battery is known as direct current because it never changes direction.

Let's see how mechanical energy of motion gets changed to electrical energy, and in the process find out how the resulting current turns out to be ac rather than dc. While the discovery was made by Faraday, the explanation depends on Maxwell's theory, which showed that a flow of current is inescapably associated with any motion of a conductor within a magnetic field. This is not to say that magnetism and motion create current—just that they always occur together. Provide any two, and you'll get the third.

If the magnetic field's strength is held constant, then the faster the motion the greater the current. If the field's polarity is reversed, the direction of current will reverse.

Imagine, if you will, a wire conductor stretched flat across a large tabletop which is covered by a sheet of paper to provide us with a smooth surface. Figure 2-6 gives the idea. Imagine also a small but powerful magnet which we can move across the table.

If we start with the magnet at one end of the table, well away from the wire, and move the magnet at a constant speed along the length of the table so that it passes over the wire and then moves away from it, and at all times we monitor the current in the wire, we will find that the instant the magnet begins moving a small current flows.

As the magnet approaches the wire, the current becomes greater, reaching a maximum as the magnet crosses the wire.

As soon as the magnet crosses the wire, the current begins to drop, becoming smaller as the magnet moves farther away.

These changes are due partially to the fact that the strength of the magnet's field depends in part upon the distance from magnet to wire, so that when the magnet is far away the field strength at the wire is less than when the magnet is next to the wire, and partly on the difference in velocity due to the varying distance.

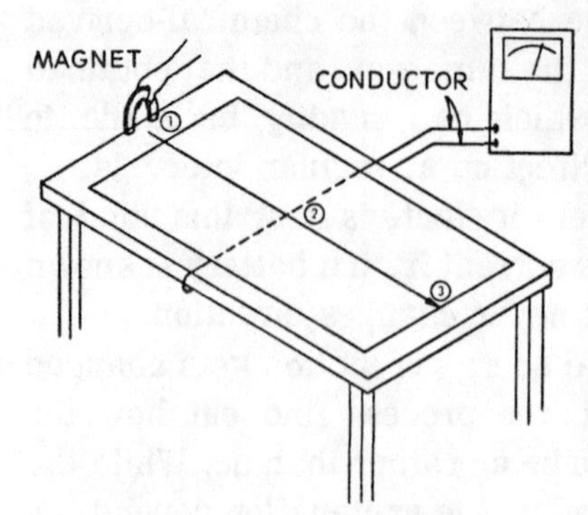

Fig. 2-6. This sketch shows the principles of converting mechanical motion to electromagnetic energy. Magnet starts from point 1, moving over conductor at 2 and continuing to 3. Motion of magnetic field is accompanied by changing electric field around conductor, which shows up as flow of current through conducting loop and meter. In practice, magnet and conductor are arranged as wheels so that motion can be continuous.

That is, if the table were 200 inches long so that the starting point was 100 inches from the wire, and we moved the magnet at a rock-steady velocity of one inch per second, then the first inch of movement (from 100 to 99 inches distance from the wire) would take place in the first second. This is about 1 percent of the total distance from magnet to wire, so the effective velocity as seen from the wire is low.

However, when the magnet has reached a point 10 inches from the wire, the next second carries it to a distance of 9 inches, which is 10 percent of the total distance, or 10 times the effective velocity of the first example.

When the distance is down to 2 inches, the 1 inch moved in one second is half the total, and the percentage figure becomes astronomical.

Actually, our example is totally impractical; but it shows the general principles involved, and is an experiment you can set up to demonstrate the principle of a generator. In practical applications, rotary motion is the easiest kind to handle, so let's use a specially constructed stationary magnet with a moving rotary conductor, as shown in Fig. 2-7. We can spin the shaft with a waterwheel, steam turbine, diesel engine, or even a handcrank. Again, we'll use a single loop of wire connected to a sensitive galvanometer to see how things happen. In the sketch, no shaft is shown. Let us assume the shaft is an invisible clear plastic rod onto which two slip rings are attached—one black and one white. The loop begins at the white slip ring, enters the area between the magnet's poles, then returns and is connected to the black ring. To simplify the explanation, let us paint the conductor, too—half the loop is black, and the other half is white, and the loop's terminal points are the rings of like color.

In sketch 1 of the figure, the white conductor moves to the left while the black moves to the right. The frozen position of the conductor and magnet shown in this first sketch depicts the zero or no-output condition. As the loop continues in a clockwise direction, the voltage increases due to the loop cutting more lines of force in a given amount of time. In position 2, when the black portion of the loop is lined up perfectly with the north pole of the magnet and the white portion is aligned with the south pole, the output voltage is at its maximum (indicated by the circled 2 above the sine wave).

As the loop continues to rotate to the position shown in 3, the output drops to zero again. Note that positions 1 and 3 are opposite, but the output is zero in both cases. In the sine wave, this situation is shown graphically at point 3.

If the loop is turned further through an angle of 90 degrees in the same direction (sketch 4), it will again be cutting lines of force at a maximum rate; however, the output voltage will be reversed in polarity. Note that the white conductor portion is

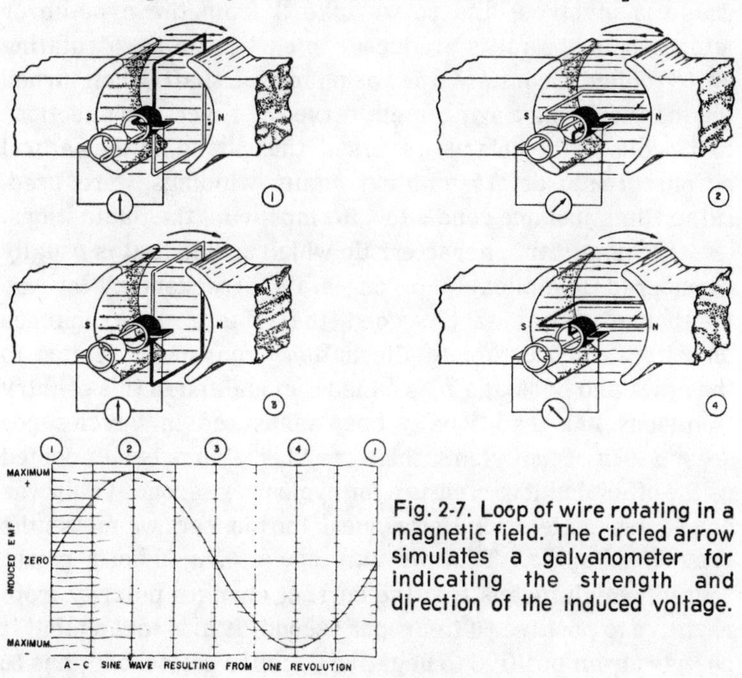

Fig. 2-7. Loop of wire rotating in a magnetic field. The circled arrow simulates a galvanometer for indicating the strength and direction of the induced voltage.

now adjacent to the north pole and the black portion is adjacent to the south. The reversal in polarity occurs because of the change in the direction in which the conductor passes through the magnetic field.

Regardless of the speed of rotation of the shaft of this basic ac generator, the output will remain a sine wave such as that depicted in the figure. With a change in the generator's rpm, there is a corresponding change in the magnitude and frequency of the output voltage. But the output maintains the sinusoidal character shown.

In the circuit of which our wire is a part, the result is a continual reversal of current flow as the loop passes by first one pole and then the other. At the instants when the loop is equidistant from both poles, current flow is zero because the opposing fields of the two poles cancel one another. In between these zero crossings, current flow depends upon the field strength and conductor velocity, and so builds up to a peak as the conductor passes the poles. Polarity of current depends upon which pole is inducing the voltage; and so we have an alternating current.

Nowadays, ac is the most useful kind where heavy power usage is involved, and so we take it from the dynamo or alternator just as it is produced. In earlier times, a rotating switch called a **commutator** was put on the shaft which turned the magnet. This arrangement served to reverse connections to the coils as the polarity reversed, thus changing the natural ac output into dc. In addition, many windings were used, rather than just one conductor, to smooth out the fluctuations.

One important characteristic which ac has that is usually considered inapplicable to dc is its "frequency," or the number of times each second that the current changes direction. Each complete alternation, from one direction to the other and back again, is called a **cycle**, and in this country frequency has traditionally been measured in "cycles per second." In recent years, the term "hertz" has been adopted as the official unit of measure for cycles per second. (This was done to honor Heinrich Hertz, one of the pioneers we met in the previous chapter.) Thus we may speak of a 60-hertz power source, which means that the current changes polarity from negative to positive 60 times per second. It also means that it changes from positive to negative 60 times per second, has 60 positive peaks per second, and has 60 negative peaks per second, since a single cycle includes the change from negative to positive, the positive peak which follows, the change from positive to negative, the subsequent negative peak, and ends at the next change from negative to positive. While we picked the negative-to-positive current reversal as the reference point in that description, actually a cycle may be measured from any reference point to the next occurrence of the same conditions.

This leads directly to a related characteristic of ac—its **phase**. If you have two different ac waveforms, both of the same frequency but not crossing zero at the same time (Fig. 2-8), a phase difference exists between the two. That is, phase is a measure of timing between two signals of a given frequency. It is meaningless to speak of phase relations between signals of different frequency, because they have no reference points in common. Ordinary household ac is single-phase, meaning that only one waveform exists in the circuit, but industrial power lines often involve three-phase circuits, usually fed over four or five wires.

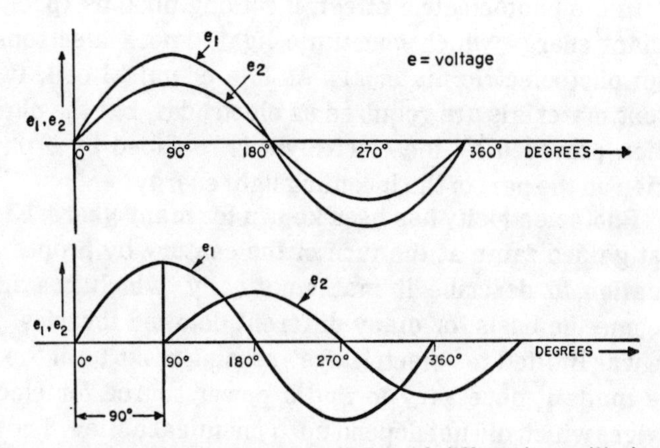

Fig. 2-8. In the top sketch, two ac voltages of different amplitudes are shown in phase—that is, both voltages are of the same frequency and cross zero at precisely the same instant. The lower sketch shows two ac voltages that are out of phase with each other. A complete cycle—zero crossing, positive peak, zero crossing, negative peak, and zero crossing—may be divided into electrical **degrees**, the total number of which equal the number of degrees in a full circle: 360. Thus, the bottom sketch shows the lower amplitude voltage as being 90 degrees behind the high-amplitude voltage. This phase relationship may be expressed in either of two ways: The high-amplitude voltage **leads** the low-amplitude voltage by 90 degrees; or the low-amplitude voltage **lags** the high-amplitude voltage by 90 degrees.

While most people consider the term "frequency" as not being applicable to dc, it's perfectly proper to think of dc as being simply ac of zero frequency. That is, its polarity goes through zero reversals, or simply never changes. Thinking of the two in this manner brings out the essential unity of all the basic concepts of electronics. Those which turn out not to apply to dc are automatically canceled by the frequency of zero, which cannot be done if you think of ac and dc as being, somehow, entirely different types of activity.

MISCELLANEOUS ENERGY SOURCES

In addition to chemical action and mechanical conversion, a number of less significant means of obtaining electromagnetic energy exist. Among them are the action of light (photoelectricity), pressure (piezoelectricity), heat (thermoelectricity), and combinations of these.

All of these are similar in some respects to chemical sources—the release of electrical energy is a byproduct of other actions which result in an imbalance of electrons.

In the photoelectric effect, incoming photons (packets of radiant energy which constitute light) knock electrons free from photoelectric materials. As in a chemical cell, two different materials are required as electrodes, but the chemical action provided by the electrolyte is replaced by a physical action on the part of the incoming light energy.

Photoelectricity has been known for many years. Einstein first gained fame at the turn of the century by proposing an equation to describe it mathematically, which has in turn become the basis for many different detailed theories. Early use was limited to "electric eye" sensing tasks until the space age made it necessary to find a power source for electrical devices which did not depend upon chemical action. The result was the "photovoltaic" or solar cell, providing usable power by direct conversion of the sun's light into electricity.

A typical present-day photocell of this type is a flat metal plate about 1/16 inch thick, with one dark surface which contains silicon, the photoelectric material. Size may range from that of a postage stamp to several square inches.

The maximum voltage obtainable from a photocell is determined by the materials used, just as in chemical cells. As technology improves, so increases the output potential and capacity per cell. Some space-system photovoltaic cells produce a large fraction of a volt at several hundredths of an ampere. Practical power extraction requires a buildup of voltage by connecting many cells in series, as well as a buildup of current capability by connecting many cells in parallel. Huge solar panels are the result of thousands of cells placed in series-parallel configurations.

The piezoelectric effect has not been used to provide power sources, but is important in electronics because it provides a widely used means of measuring motion, and the most commonly employed method to control the frequency of a radio signal.

Piezoelectricity is due to an unusual crystal structure found in several materials. Among these materials are quartz, Rochelle salt, and several man-made ceramics. When a piezoelectric substance is bent, pressed, or twisted, an electric field appears across its faces as a direct result of the motion of its structure.

The common "crystal" or "ceramic" phonograph cartridge makes use of this effect to convert the motion of the phonograph stylus into an electrical signal which duplicates the waveform of the original sounds. A crystal or ceramic microphone uses the same effect. Here, a large diaphragm is connected to the crystal. Sound waves vibrate the diaphragm, which distorts the attached crystal enough to produce an electrical analog of the sound.

In radio, specially ground crystals of quartz are used as frequency standards. The frequency is established by the mechanical resonance of the crystalline plate, and is converted to an electrical waveform by the piezoelectric effect.

Like piezoelectricity, thermoelectricity has been limited in application to essentially a "sensing" role—except for its application to space exploration in the form of the "fuel cell." In thermoelectricity, electrons are freed by heat energy. Two dissimilar metals are used, usually bismuth or antimony and copper. As in chemical cells or photocells, the voltage is determined by the materials used for the electrodes; the current is governed by electrode area.

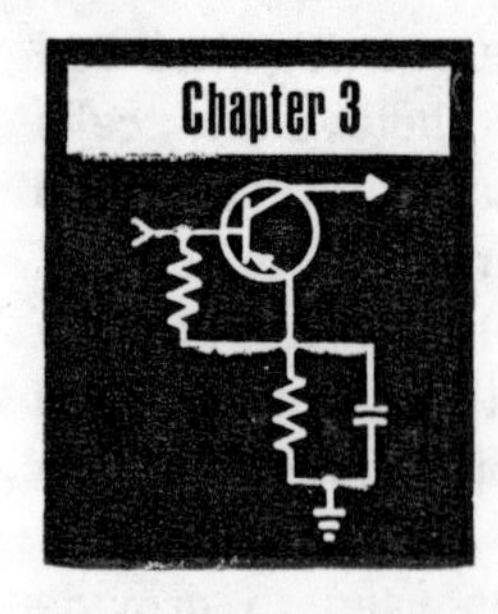

Passive Circuit Elements

In electronics, nothing at all can happen until you have a circuit for it to happen in. Chapter 4 will get into the details of just what constitutes a simple circuit and what goes on there, but before we can put together even the simplest of circuits we must have something to put it together with. That's what this chapter is all about.

In order to get started at all, though, we must break into the circle of definitions and create an entry point. We'll do that by defining a circuit as **any complete path in which electromagnetic energy can flow.** A circuit usually includes an energy **source** and one or more energy **sinks.** The energy **source** provides the energy which flows in the circuit, and the **sink** dissipates the energy.

With "circuit" defined, we can now define a **circuit element** in much more simple terms: **any specific component which may be included as a part of a circuit.**

The most basic circuit element of all is a conductor, which permits the other elements of the circuit to be connected together to form the circuit itself. The conductor is so basic that most discussions tend to take it for granted. Since we've already made the acquaintance of the conductor back in Chapter 1, we'll merely note here that it is in fact a circuit element, and move on to the more complicated passive elements which may be included in circuits.

In this chapter, incidentally, we will consider only **passive** elements, which are those elements which neither add to nor detract from the amount of energy present in the circuit. (Dissipation of energy in the form of heat is not considered here as detracting from the amount of energy, since the heat is still associated with the circuit.) **Active** elements, which modify the amount of energy involved, are discussed in Chapter 5.

The general classes of passive element which we'll examine here are the **switch**, the **resistor**, the **capacitor**, **inductors**, and **indicating devices**. Ready? Let's get on with it.

THE SWITCH

While the conductor, as we already noted, must be the most basic and simplest circuit element of them all, the switch runs it a close second. The purpose of the switch is to make or break a circuit. Occasionally, a number of simple switches may be combined into a multiple unit which is still called a switch although it affects many circuits, but each of its basic units still either makes or breaks a single circuit.

The oldest type of switch, and one which still finds use in restricted applications, is the classic "knife" switch shown in the terribly impractical circuit of Fig. 3-1. When the pivotal bar is pressed down, the circuit is "complete," and current flows. In the case shown—assuming the battery is in a full state of charge—when the switch closes, the full capacity of the battery is released; it short circuits, causing the conductors to burn, the meter to be damaged, and perhaps the battery to be destroyed from overheating. To make this circuit practical, a load (such as a lamp) must be inserted in the circuit to resist the flow of electrons somewhat.

Any switch can be characterized by the number of poles it has (that is, the number of circuits involved), and the number

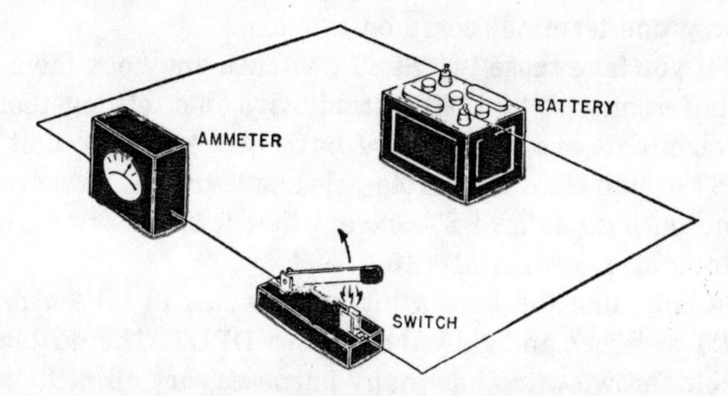

Fig. 3-1. The basic circuit. When the switch closes, the battery delivers full current into the circuit. Since an ammeter typically offers little resistance to current flow, a complete circuit like this won't be functional long. Either the battery will run down quickly, the wires will burn up, or the ammeter will be blown apart by the furious electron traffic. If a voltmeter were substituted for the ammeter, the circuit would be sound. (A voltmeter offers substantial resistance to current flow.)

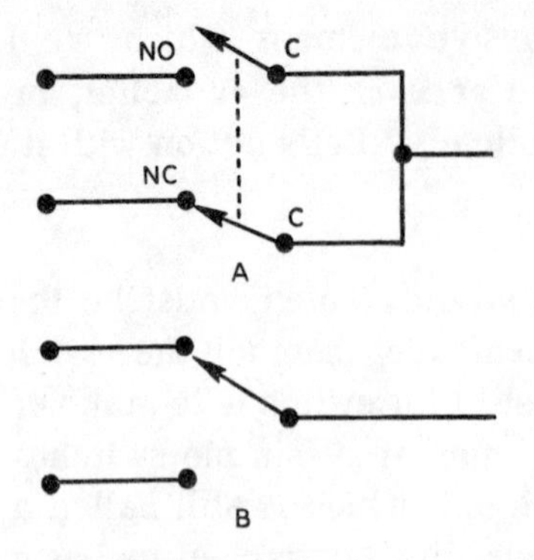

Fig. 3-2. Action of SPDT switch (bottom) is identical to that of ganged SPST switches shown at top. This typifies manner in which more complicated circuits are built up from simpler circuits and circuit actions, which continues throughout electronics. **The most complex computer imaginable can be reduced to a large number of simple switches.** NO = normally open; NC = normally closed.

of "throws" it can make. The ordinary light switch is a single-pole single-throw switch, abbreviated SPST. This means that it controls a single circuit (single pole), and has only two positions, on and off (single throw).

A single-pole double-throw (SPDT) switch (Fig. 3-2) would still control only a single circuit, but would have three terminals instead of two. In one position, the circuit between terminals NO and C would be made and that between NC and C broken, and in the other position the NC-C circuit would be made and the NO-C circuit broken. If the NC terminal is left out of things, then the SPDT switch acts just like a SPST for the other two terminals. Similarly, if the NO terminal is ignored, it's an SPST for the other two. Thus, you can think of an SPDT switch as being like two SPST switches interconnected so that when one is on, the other is off, and having one terminal common to each.

If you take those two SPST switches and hook them up a bit differently, so that both circuits are on or off together and no terminals are shared, you have converted the unit to a DPST switch (that's a double-pole single-throw). It serves the same purpose as a SPST, except that it breaks two circuits instead of one when operated.

Continuing the idea a bit more, you can take a pair of SPDT switches and gang them into a DPDT. The double-pole double-throw switch has many purposes, not all of them obvious. For instance, if its two arms are hooked to the positive and negative terminals of a battery, and the other four terminals are cross-connected, you can use it for polarity reversal. If you connect three of the four outer arms to the positive terminal of a battery, leave the fourth disconnected, connect one arm to an "accessory" outlet, and the other arm to an ignition coil circuit, you have the typical automobile ignition switch circuit in which one position provides full operation and the other powers only electrical accessories while no power is available to the ignition coil.

This example introduces another feature of many switches. The auto ignition switch has a "center-off" feature, so that an "off" position is available in addition to the two throws provided by the double-throw feature. While this can also be described as "triple throw" with the center terminals left unused, most toggle switches can be obtained with "center off" action.

Still a fourth feature entering into the description of many switch types is the difference between **positive** and **momentary** action. The light switch has **positive** action; once tripped, it remains in position until tripped again. A doorbell pushbutton has **momentary** action; it keeps the circuit closed only as long as it is pressed and opens the circuit when released.

Not all pushbuttons are momentary action, of course. Many alternate-action pushbuttons are used in industrial electronics. The first time one of them is pressed, it closes the associated switch contacts. The next time, the contacts are opened. This continues, with the action alternating from closing to opening on every other operation.

Most switches make use of contact points, which simply wipe two conducting surfaces together firmly to establish the circuit in the closed position, and move them apart to open the circuit.

Selector Switches

Almost all selector switches are multiple-throw devices, although it is possible to use SPST switches arranged to short out the component to be selected. When the switch is closed, in such an arrangement, the component is out of the circuit; when the switch is open, the component is "selected."

The most common selector switch (Fig. 3-3) is the **rotary** type. Many of them are multipole or "ganged" switches. The bandswitch in a radio communications receiver, which selects one of a number of tuned circuits, may have nearly a dozen poles, each having five or six positions. While this **could** be called a "12P6T" switch (as in SPST or DPDT), it normally isn't.

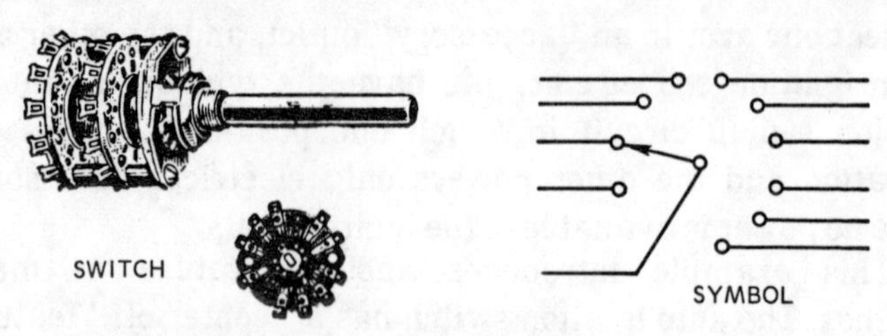

Fig. 3-3. Rotary selector switch allows its wiper contact to close circuit to any of several other contacts. Schematic symbol, right, indicates action clearly. Many "decks" can be ganged to produce complex switching structures, as in radio receiver bandswitches or multimeter range selectors. Switch at top left, for example, is **"two-gang."**

A frequent alternative to a rotary switch for use as a component selector is the multiple pushbutton switch. In some cases the rotary's requirement that each component, in turn, be selected until the desired one is reached (a process which occurs as the knob is turned to make the selection) is unacceptable. When this is the case, the pushbutton (which jumps directly from the previous selection to the new one, without passing through intermediate points on the way) is used.

In either case, the selector switch has an additional characteristic which is not usually specified on power-control switches. This is the relationship between actions at the various contacts. For instance, a switch may be described as having a **make-before-break** action. This means that the new circuit will be made or closed, before the old one is broken or opened. Similarly, **break-before-make** action means that the old circuit is broken before the new one is made. The make-before-break action is sometimes described as **shorting**, while break-before-make may be called **nonshorting**, but the term "shorting" has also been used for a type of selector switch action which never breaks old contacts, simply adding more and more new ones. This action is sometimes desirable.

It's important, in selecting a switch for use in a new circuit, or when replacing an existing one, to make sure that all the proper characteristics are kept in mind. For instance, the difference between make-before-break and break-before-make action is minute. Electrically, it makes no difference which is used so long as the switch is never operated with power applied to the circuit. But if a break-before-make switch is used as the function selector in an audio amplifier, to switch from radio to phono to tape for input, then every time the switch is operated while power is on the speakers will emit a loud pop, because the amplifier's input circuit will be left open momentarily as the switch operates. Similarly, using a make-before-break switch as a power selector would connect both power sources together when operated, probably melting the switch.

A Special Safety Switch

While we're examining switches, there's one special type which must be looked at. This is an automatic switch, good for only one operation, which destroys itself when it operates.

It might appear that only an idiot would make use of any circuit element which had to self-destruct to accomplish its purpose, but good circuit design not only permits, but demands, inclusion of the safety switch known as a fuse.

The whole purpose of the fuse is to switch off power to the circuit before anything other than the fuse can be harmed. It's the old principle of a chain being only as strong as its weakest link. Long ago, engineers discovered that the best way to protect everything else was to make one link deliberately weakest, so that it would fail first. The very name "fuse" is an abbreviation for "fusible link," based directly on the chain analogy.

A fuse is designed to carry its rated current and act as a conductor. When current increases beyond the rated value, the fuse melts, acting as an SPST switch and opening the circuit.

The ordinary fuse melts within 10 seconds when current rises to twice its rated value, and in less time if current is greater. It will last indefinitely at current values up to the rated value.

Slow-blow fuses are also available which will carry up to 10 times rated current for brief periods of time. They are intended for use in circuits which require high surges of current when turned on, but much lower values for extended operations.

THE RESISTOR

Back when we were defining current and its unit, the ampere, we mentioned that one definition of the ampere would be the amount of current forced through a resistance of one ohm by a potential of one volt. The definition was not very useful at that point, because we had not defined either the volt or the idea of resistance.

Resistance is the property which tends to limit current flow in a circuit. It's due to the interaction between the free electrons and their "bound" brethren in any conductor at normal temperatures. (At very low temperatures, all resistance disappears from some materials. Bell Labs has successfully passed large currents through a thin-film alloy of niobium and germanium at a temperature of 418 degrees below zero **without resistance losses.** But this **superconductivity** is beyond our immediate interest.)

The unit of resistance is the ohm, in honor of the man who first formulated the relationship between resistance, voltage, and current. **Ohm's law** is so fundamental to all theory of electronics that it may come as a surprise to learn that Ohm created it as a byproduct of his studies into the effects of heat upon resistance, and in his original publication did not even express it in the form it bears today. The modern version of the law, that voltage is equal to current times resistance, was invented by Maxwell, who included it as one of the 20 equations of his dynamical theory.

The interaction between free and bound electrons which we know under the name of **resistance** causes some of the energy originally available for free electrons' use to be dissipated as heat, and is the major factor limiting current flow in any specific material. The amount of interaction depends both upon the material itself, and upon the physical size of the conductor involved. Different materials have different ratios of free to bound electrons, and the larger the conductor the more electrons are available within any specific portion of it. Materials with a high ratio of free electrons to bound electrons we call good conductors, and this property is described also as that of having low resistance. The better the conductor, the lower the **resistance**.

Insulators have very high resistance, of course, in order to insulate. The resistance of a good insulator is so astronomically high (in the order of billions of ohms) that most of us tend to think of it as infinite, but the resistance of just about any material can be measured with sensitive enough equipment.

In the gap between good conductors and good insulators we find materials having too much resistance to be considered good conductors, yet not enough to be classed as even poor insulators. These materials offer us a means of packaging resistance to use wherever we may need it in an electrical circuit. One such material is carbon. Another is the nickel-chromium alloy we know as nichrome.

It may sound strange to discuss the idea of "packaging" resistance for use wherever it may be required, but the resistor, a true packaged resistance, is one of the most useful circuit elements around. Much of its usefulness derives from its ability to control either current or voltage in a circuit, but it has other applications as well.

One of the most prominent bits of packaged resistance is the incandescent lamp, or ordinary light bulb. This is, in essence, a tungsten resistor (Fig. 3-4) operated under conditions which cause it to glow white-hot. This makes it give off light.

It may still be used as a resistor. When electronics replacement parts were almost impossible to obtain during World War II, light bulbs (which were not rationed) were used by ingenious radio repairmen to serve as replacements for power resistors. They have also been used to drop power to soldering irons—to keep the iron warm but not at full heat.

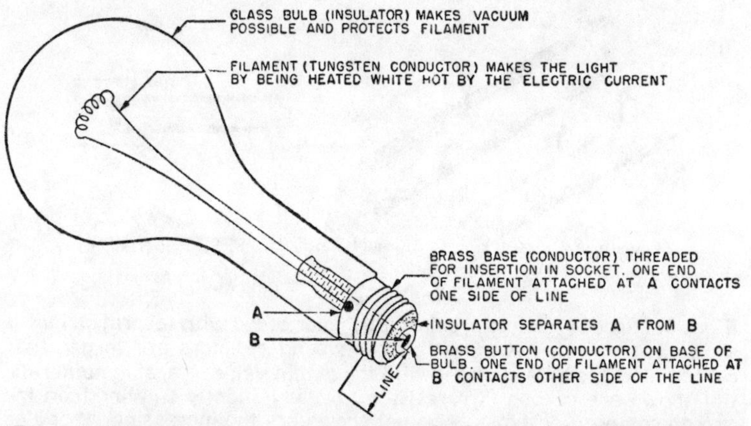

Fig. 3-4. The ordinary incandescent lamp bulb is a packaged resistance. It is comprised of a number of insulating and conducting elements, as shown.

Not so common, but still frequently encountered, is another everyday use of the resistor—the electric heating element. Electric stoves, ovens, coffeepots, irons, toasters, and so forth all employ heating elements which are nothing but resistors. These resistors have greater area for their resistance than do light bulbs, so that they do not get white hot. As a result, all the energy which they dissipate is turned into heat and none is lost lighting up the countryside. Also, the element lasts longer. Again, light bulbs produce considerable heat, and can be used as substitutes for other heating elements.

In most electronics applications we need to neither light things up nor heat them. Our uses of packaged resistance revolve, instead, on its capability of controlling power. For these applications, we need a wide variety of resistance values in a selection of dissipation ratings, and that's what we have available under the general name of **resistors**. We can take our pick from several types of composition, tolerances ranging from 1 to 20 percent, power ratings from a fraction of a watt to 100 watts, and we may obtain fixed, variable, and adjustable resistors. Let's look at these in a little more detail.

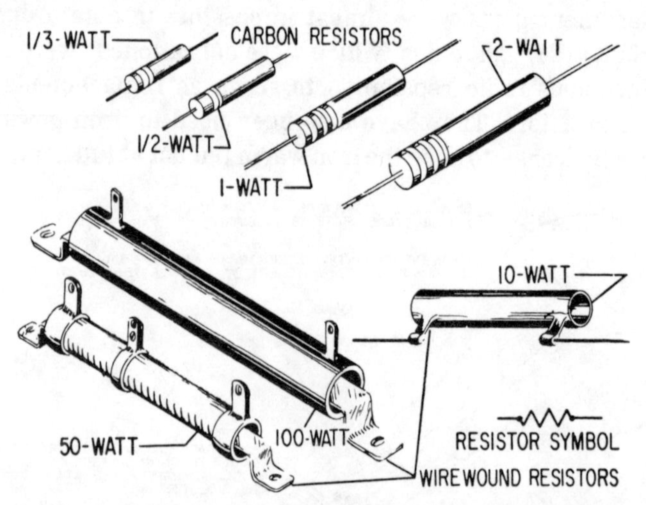

Fig. 3-5. Carbon resistors are normally color coded with several stripes to show resistor value. Ordinarily, wirewound resistors are larger than carbon, and are thus stamped with the actual value in arabic numerals. Not shown are carbon film resistors, usually slightly smaller than the carbon composition types. Also not shown are the increasingly popular tenth-watt, eighth-watt, and quarter-watt sizes typically used with integrated circuitry.

Typical Resistor Construction

The most common type of resistor in electronics is probably the half-watt carbon (composition or film) resistor. Both types look like a short tube with wire leads extending from each end; they're available in resistance values ranging from less than an ohm up to tens of megohms (millions of ohms). The resistance is provided by either a composition material containing carbon, or a deposited film of carbon over a nonconductor. The relative sizes and configurations of the more popular resistors are shown in Fig. 3-5.

Similar in appearance to the fixed composition resistor is the fixed insulated wirewound resistor. The difference is that in the wirewound unit, the resistance element is a wire rather than a composition compound. This limits the resistance range to 470 ohms maximum in the half-watt size, and 3300 ohms maximum in the 2-watt variety.

Both carbon and wirewound units of low wattage ratings are usually marked as to value by a code of color bands around the resistor body. Either three or four stripes may be used. The stripe nearest the resistor end (Fig. 3-6) gives the first digit of the value, the next one gives the second digit, the third tells the number of zeros following the two digits, and the fourth stripe, if present, denotes the tolerance. The standard color code is used, in which black represents zero, brown is one, red is two, orange is three, yellow means four, green is

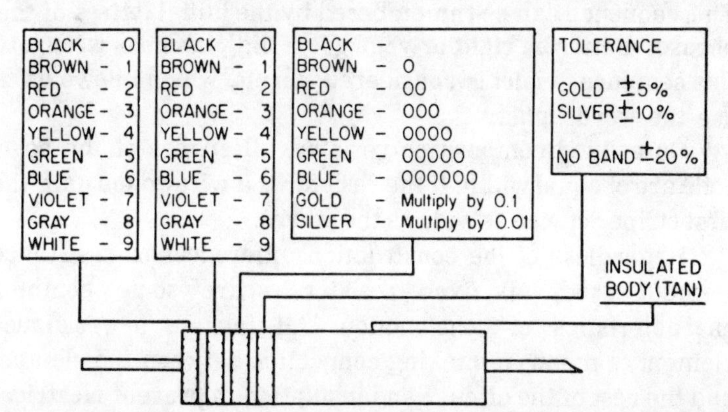

Fig. 3-6. The resistor color code tells you the value in ohms, while the size of the resistor tells you the amount of power it will safely dissipate as heat. To read the resistance value, hold resistor so that group of color bands is at left side of resistor, then convert each of the first two colors to a numeral (as shown). The third band's numeral simply tells you the number of zeros to add to the first two digits.

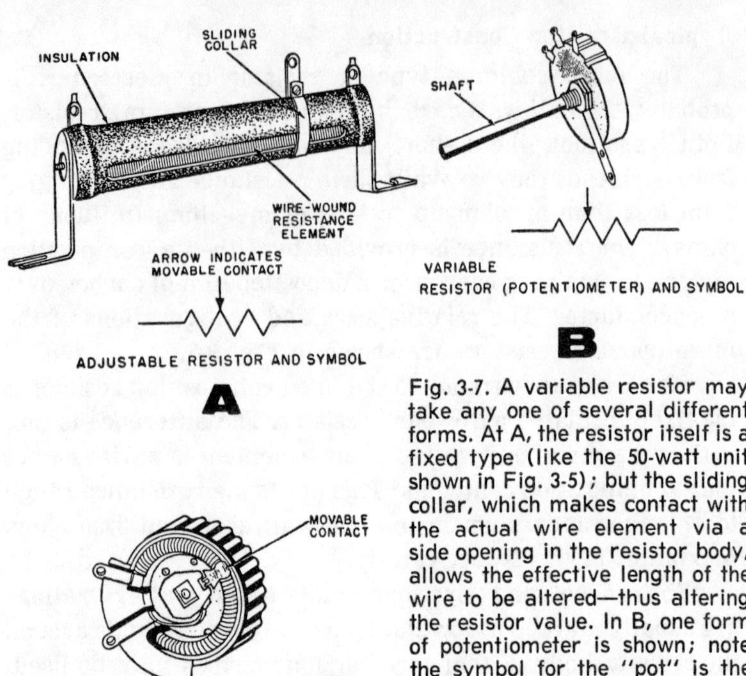

Fig. 3-7. A variable resistor may take any one of several different forms. At A, the resistor itself is a fixed type (like the 50-watt unit shown in Fig. 3-5); but the sliding collar, which makes contact with the actual wire element via a side opening in the resistor body, allows the effective length of the wire to be altered—thus altering the resistor value. In B, one form of potentiometer is shown; note the symbol for the "pot" is the same as that of the variable resistor. The rheostat (C) is characterized by the appearance of relatively large amounts of power in its control applications.

five, blue is six, violet is seven, gray is eight, and white is nine. This sequence can be remembered by the initial letters of this phrase: **Better be right or your great big plan goes wrong**. In this sentence, **violet** is considered **purple**, which allows us to use the word "plan."

On a fixed composition resistor, all stripes of the color code are of equal width. If the resistor is a wirewound unit, the first stripe is twice as wide as the others.

Regardless of the construction differences or resistance element used, all fixed resistors share some common characteristics of construction. All include a resistance element, a means of making connection between the element and the rest of the circuit, and insulation to prevent electrical contact except where desired.

Typical Variable Resistors

Not all resistors are fixed in value. One of the most common control devices in electronics is the variable resistor

(Fig. 3-7); the most familiar form of it is probably the radio "volume control."

The volume control, more accurately called a **potentiometer** (often abbreviated to "pot"), has only one major functional difference from a fixed resistor. The potentiometer has, in addition to the connections at each end of the resistance element, a third movable connection to the element, which may make contact anywhere from one end to the other. If one of the end terminals is ignored and connection made instead to the movable sliding arm or **wiper** contact, the result is a resistor which can be varied from a value of zero ohms when the arm is at the end of the element connected to the other terminal, to the maximum value when the arm is at the other end of the element.

If both ends of the element are connected to a circuit, then the arm offers a means of tapping off only a part of the signal flowing through the element. This is the normal "volume control" application, and we'll look at it in more detail in the next chapter.

In most control applications, the position of the arm or wiper contact is determined by a shaft which can be turned. The typical potentiometer has a shaft capable of turning through 270 or 300 degrees of rotation, less than a full circle. For industrial purposes, 10-turn potentiometers are often used. These require 10 full turns of the shaft to move the arm from one extreme to the other, and so offer greater precision in control. Some miniature units use a threaded-rod adjustment which achieves similar results.

When used to control power, the variable resistor normally uses the two-terminal configuration and is known as a **rheostat**.

In addition to the shaft-operated control, resistors which resemble power resistors but which have a movable tap contact are available. These are known as **adjustable** resistors. The upper resistor shown in Fig. 3-7 is an example; note the sliding tap girdling the resistor. The major difference between an adjustable and a variable resistor is that the variable may be changed in value safely with power applied; changing the setting of an adjustable resistor requires that power be removed before the adjustment is touched, however.

Both variable and adjustable resistors are available in the same range of materials as fixed units.

The Importance of Power Ratings

While resistors are used in electronics primarily to control circuit actions, they can accomplish their purpose only by converting electrical energy into thermal energy, or heat Such a conversion is the only thing a true resistance does. Many effects other than true resistance can also cause energy to be "lost" from a circuit, and these effects are usually called "resistance" and measured in ohms because of their similarity to actual resistance action; but despite this, the only action of a true resistor is to convert electrical energy into heat.

The heat thus produced is measured in terms of watts. The wattage rating of a resistor is usually based on the amount of power input which will cause the resistor's temperature to rise from 40 to 110 degrees (Celsius) in free air. That 110-degree temperature is 10 percent higher than the boiling point of water, or about 233 degrees Fahrenheit. If such high temperatures are not acceptable, or if the resistor is in a confined space where air cannot circulate around it, or if the starting temperature is higher than 40°C, a higher power rating should be used.

In practice, the power rating of any resistor should be considered as only a **maximum** worst-case rating. The usual rule followed by equipment designers for choice of a resistor's power rating is to calculate the power which the resistor is actually going to convert into heat, double this value, and take the next higher standard unit. Resistors smaller than the half-watt types are usually found only in spots where physical size is important, because they have traditionally been more expensive.

The typical volume-control potentiometer usually bears a half-watt rating, although in many cases even this is too much. The reason is that such controls are often used in sensitive audio circuits, and letting them dissipate measurable amounts of power introduces excessive noise. Power rheostats and adjustable resistors follow the same range of standard values as power resistors.

Power ratings for carbon film resistors are based on continuous operation, and may also be affected by peak values attained during pulses, because the film is physically small and cannot average its temperature as well as the thicker composition rods or wire of other types.

Tolerances and Standard Value Series

All through this discussion we have been using the word **tolerance** without bothering to explain what it means. The tolerance is an important part of the value specification, not only for resistors but for all other circuit elements. Absolute accuracy is not achievable in any physical measurement or manufacturing; the best that can be done is to bring error within defined acceptable limits. The tolerance defines what these limits may be.

That is, a 100-ohm resistor with a tolerance of 20 percent might actually have any value between 80 (100−20) and 120 (100+20) ohms. A 100-ohm unit with a tolerance of 5 percent would be between the limits of 95 (100−5) and 105 (100+5) ohms.

Standard tolerances for resistors are 20, 10, 5, 2, 1, 0.5, and 0.1 percent. In the color code, 20 percent tolerance is indicated by absence of the tolerance stripe. A silver band means 10 percent, and a gold band is 5 percent. Fixed composition resistors normally are not available in tolerances below 5 percent, but the color code specifies that 2 percent would be indicated by a red band and 1 percent by a brown one.

The concept of tolerance also sets the standard value series employed for most electronics components. It simply would not make sense to furnish both a 90-ohm and a 100-ohm resistor in 20 percent tolerance, since the 100-ohm unit would range from 80 to 120 ohms and the 90-ohm one would span from 72 to 108 ohms. For a large part of the allowable range, the two units would be interchangeable.

In consequence, a set of standard value steps based on the idea of making the tolerance limits just meet has been developed. For 20 percent tolerances, values of 10, 15, 22, 33, 47, and 68 (and multiples by 10) are used. The upper limit of tolerance on a 100-ohm unit is 120 ohms, and the lower limit for a 150-ohm unit (the next step in the series) is also 120 ohms.

Thus any possible value comes within 20 percent of one of the six steps.

When the tolerance is reduced to 10 percent, the six steps of the previous series are retained, and the midpoints are added, so that the resulting 12-step series runs 10, 12, 15, 18, 22, 27, 33, 39, 47, 56, 68, and 82.

At 5 percent, the midpoints of the 10-percent steps are added to produce a 24-step series of 10, 11, 12, 13, 15, 16, 18, 20, 22, 24, 27, 30, 33, 36, 39, 43, 47, 51, 56, 62, 68, 75, 82, and 91.

While the use of a 5100-ohm resistor in place of a 5000-ohm unit may be a trifle confusing at first, it eliminates many unnecessary values from the manufacturers' stocks and greatly simplifies repair work. Whenever a more precise value is needed, the designer can move to the 1 percent or 0.1 percent units to get it—and if he doesn't need such accuracy, then he doesn't need a more precise value either.

Potentiometers offer an exception to this value-series rule. They still come in values such as one-half megohm instead of the 0.47 or 0.51 megohm which would fit into the series. Power resistors are another exception, but standardization in this area is slowly advancing. "Even" values of resistors may be found in antique equipment, but not in designs more recent than the mid-1950s (when the standard was accepted industry-wide).

THE CAPACITOR

Today's capacitor is a direct descendant of the Leyden jar, the first element capable of storing electrical charge. That's still the primary function of a capacitor—the storage of charge.

The capacitor's action is explained most accurately by Maxwell's theory, which described the propagation of electromagnetic energy as a continual transformation of energy from the state of motion into storage and vice versa. If an electrical circuit is interrupted, so that the energy cannot propagate through it, the motion-compression-motion cycle must halt in the compression or storage condition.

This storage of energy occurs not in the conductor, but in the insulating medium which keeps the circuit open. It amounts to a displacement of the free electrons from their random positions to the positive side of the circuit. (See Fig.

3-8.) When all the free electrons have been swept out of the insulator, the bound electrons are no longer in electrical balance. At the negative boundary, too many electrons remain, having been forced in by the original pressure; at the positive boundary, there are too few. This displacement of the internal structure of the insulator, under applied electrical pressure or voltage, is what we call **charge** and measure in coulombs.

Charge is one of the fundamental units of electronics, being related to current by time. Another of the fundamental units is voltage or electrical pressure. Voltage and charge for any specific capacitor are related by the electrical capacity of that capacitor.

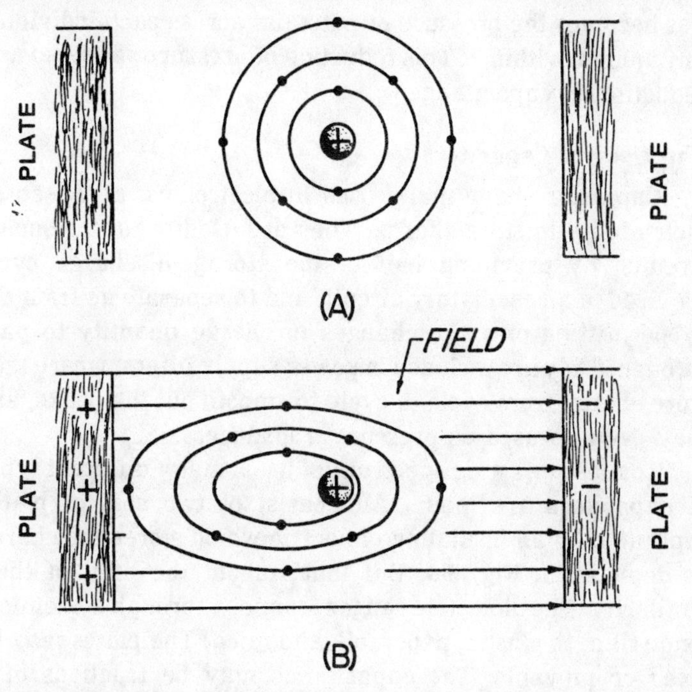

Fig. 3-8. In sketch A, there is no difference in charge placed across the plates of the capacitor (represented by shaded portions). The structure of the dielectric's atomic orbits remains undisturbed. However, if a charge is placed across the capacitor, the orbits will become elongated, as shown in B. Note that the direction of stretch is toward the positive plate, even though the positive proton in the nucleus "reaches" for the negative side. It takes energy to distort these atomic orbits; and since energy cannot be destroyed, it may be recovered when the orbits are permitted to return to their normal positions. The effect is analogous to the storage of energy in a coiled spring.

Electrical capacity is similar to, but not identical to, physical capacity. In earlier times, electrical capacity was known merely as "capacity"; capacitors were called "condensers" as well (a term still used by auto mechanics and nontechnical oldtimers in radio). Confusion led to the adoption of the word **capacitance** to mean electrical capacity; the term **condenser** was then replaced by **capacitor**.

The thicker the insulator in a capacitor, the greater is its physical volume but the smaller will be its capacitance. If area is increased, however, both volume and capacitance will increase. This seeming contradiction comes about because charge is effectively stored only at the boundaries between the insulator or **dielectric** and the conductors or **plates** of the capacitor. As the thickness of the insulator is increased, the less becomes the pressure or potential across each individual tiny volume within it. This reduction of pressure shows up as a reduction in capacitance.

The Uses of Capacitors

Capacitors have many uses in electronics, all based on their ability to store charge. They are used to tune resonant circuits, by providing half of the storage-discharge cycle required in an oscillatory circuit, and to separate ac from dc, by permitting only the changes in charge quantity to pass through. They are included in power supply filters, where they store energy from cycle to cycle to smooth out the peaks, and may even be used as pressure transducers.

To meet this wide spectrum of uses, many different types of capacitors are made. All consist of two sets of plates separated by an insulating dielectric which stores the charge as depicted in Fig. 3-9. But that's about the limit of their similarity. The dielectric ranges from air, through a chemical oxide film, to plastic, paper, oil, and mica. The plates may be fixed or movable. The capacitance may be fractions of a picofarad (millionth of a millionth of a farad) or hundreds of thousands of microfarads.

Each of the different types is best suited for only a small fraction of the myriad uses of the capacitor. Air capacitors are used primarily for tuning rf circuits, and sometimes for neutralizing power amplifiers at radio frequency. Chemical film or electrolytic capacitors find their major application as power supply filters, with a secondary use for bypassing power lines. Plastic and paper units are used for signal coupling. Oil capacitors are used at high voltage where other insulators might break down under the stress. Mica dielectric is used where the utmost in stability is necessary, as in fixed-tuned circuits and for filters. Ceramic units are used for temperature compensation, and for general bypass applications where relatively loose tolerances are acceptable. Figure 3-10 shows but a few of the many available configurations of fixed capacitors.

Early experimenters used the capacitor's ancestor, the Leyden jar, to store up charge. The energy storage capacitor performs this same function today in photoflash units, capacitor-discharge auto ignition systems, and plasma generators.

In analog computers, integrating capacitors do the same thing on a much smaller scale, to perform the mathematical operations of the integral calculus.

Capacitors are also used to measure time intervals, because their rate of accumulating charge can be made reasonably constant so that if one unit of charge is accumulated in one second, five units will accumulate in five seconds. This application finds practical use in industrial timers, radar, test equipment, but most especially in television where the timing provided by the vertical and

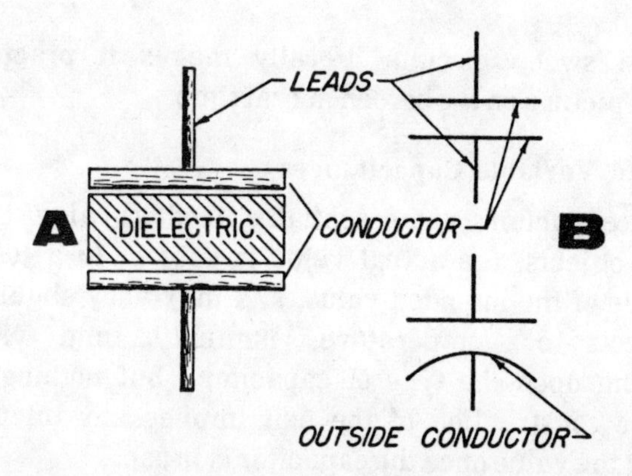

Fig. 3-9. Capacitors all have one element in common: They consist of a pair of conducting plates separated by an insulating medium known as a dielectric. Fixed-value capacitors are schematically represented by either of the two symbols shown in sketch B.

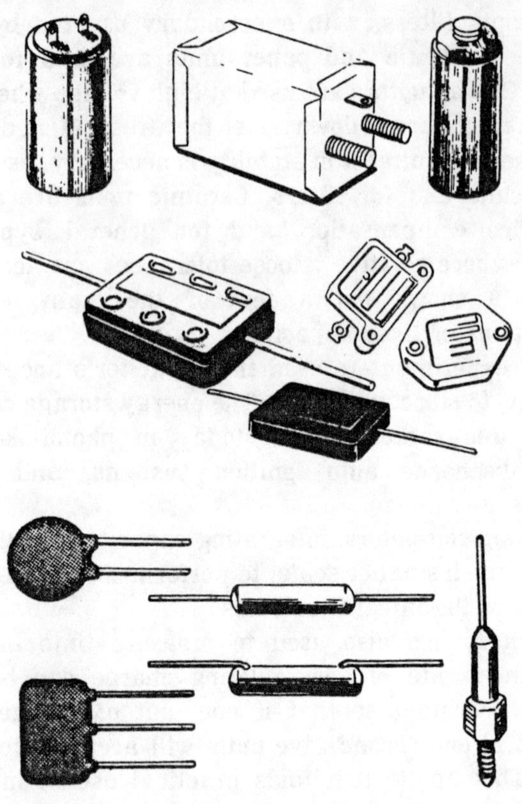

Fig. 3-10. Depending on the ultimate application, a capacitor can take any one of more than a hundred possible forms. Those shown here, not drawn to scale relative to each other, are but a sampling of the available configurations.

horizontal sweep circuits literally makes it practical to present a picture on the screen in real time.

Fixed and Variable Capacitances

Most capacitors are essentially fixed in value. Like all physical objects, the actual value is subject to a tolerance either side of the intended value, and may vary slightly with the effects of temperature, humidity, and vibration (depending upon the type of capacitor), but nothing in the design or construction of the unit implies any intention of changing the value once the capacitor is in use.

Some capacitors, however, are designed to be **variable** in value. The most common of these is the radio tuning capacitor, which is usually an air-dielectric unit with one fixed or stator set of plates and one movable or rotor set, as shown in Fig. 3-11. The rotor plates are mounted on a shaft, which can be rotated by the tuning knob.

When the shaft is in one position, the plates are fully meshed, with the rotor set fully inside the stator set. This is the maximum-capacitance position. When the shaft is rotated 180 degrees, the plates are completely unmeshed, with the rotor plates as far removed as they can be from the stator. In this position the capacitor has its minimum capacitance.

A typical range of capacitance for such a unit is 10 to 1. That is, maximum capacitance is usually about 10 times the minimum value. For many years, a maximum value of 365 picofarads (pF) was almost standard; minimum value was usually about 30 pF in this case.

While it might appear that such a capacitor would have half its differential capacitance with the shaft rotated 90 degrees (that is, half of the difference between maximum and minimum, or 165 plus 30 for the 365 pF unit), that's not necessarily or even usually the case.

The actual relationship between capacitance and shaft rotation depends upon the shapes of both sets of plates, because the capacitance depends primarily upon the area of the plates which is meshed at any point. Three major classes exist, known as **straight-line frequency, straight-line capacitance,** and **straight-line wavelength** design. The SLF

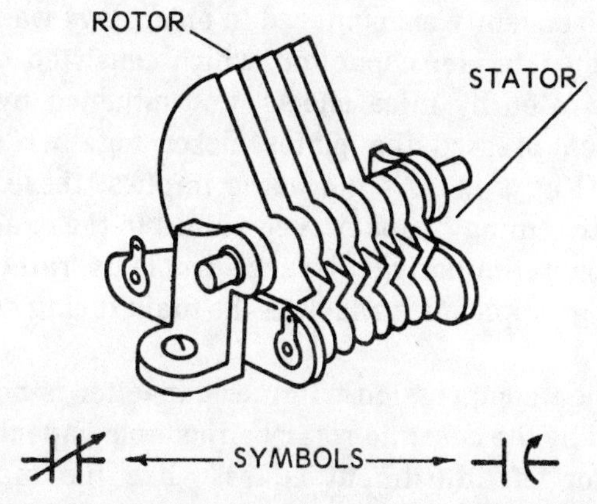

Fig. 3-11. Perhaps the most common form of variable capacitor is the air-dielectric type shown here. It consists of an immovable stator and a movable rotor. A tunable shaft allows the plates of the rotor to mesh between the stator plates without allowing any of the plates to actually come into contact with each other.

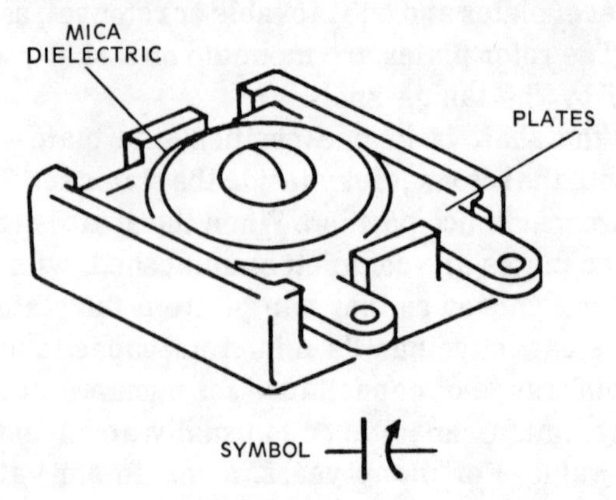

Fig. 3-12. A trimmer capacitor consists of at least two miniature plates separated by springy mica leaves. A screwdriver adjustment provides the means for tuning the device to the value required. Schematically, the trimmer may be depicted by the curved-arrow symbol, but there is no specific rule that the symbol will take this form.

design is intended to make the tuning dial read in uniform increments of frequency. The SLW design has the same goal, but for a dial calibrated in wavelength rather than frequency. Only an SLC design would give equal increments of capacitance, and few of these are made. Most tuning capacitors are aimed in the general direction of the SLF design, but do not quite reach it.

While the air-dielectric tuning capacitor is the most common variable capacitor, it is by no means the only type. Almost as frequently encountered in older days was the mica compression **trimmer** capacitor, which consisted of springy plates separated by mica sheets, and adjusted by a single screw which pressed the plates closer together as it was tightened (Fig. 3-12). As the name implies, these units are used for "trimming" capacitance values to the exact amount required, thus eliminating tolerance problems, rather than for quick-change applications such as the main tuning control of a radio.

The mica compression trimmer capacitor is now largely supplanted by the ceramic rotary adjustable capacitor. While its precision of adjustment is less (like the main tuning capacitor, the ceramic rotary goes through its complete minimum-to-maximum range in 180 degrees of rotation, while the compression units required as many as 10 turns), it holds its adjustment over a longer period of time and is less affected by humidity.

Still another type of adjustable or variable capacitor is that known as a neutralizing capacitor, intended for use in circuits which cancel out the effects of unwanted feedback in radio-frequency amplifiers. A neutralizing capacitor usually uses air as its dielectric and has only two plates. One or both of the plates is mounted on a threaded rod, so that the spacing between plates can be varied from nearly nothing at all up to several inches. This provides a wide adjustment range, together with the capability of obtaining capacitance values lower than 1 pF (which are often required in this application).

A final type of variable capacitor, the voltage-variable capacitor or **varactor**, makes use of a semiconductor junction. We'll look at this one in Chapter 5, after meeting transistors.

Types of Capacitors

Since any pair of conductors separated by an insulator forms a capacitor, any attempt to list all the possible types of capacitors would be a hopeless undertaking. Ingenious technicians have contrived capacitors from fragments of TV twinlead wire, and from short lengths of coaxial cable. A radio antenna insulated from ground forms one plate of a capacitor, with the earth itself forming the other plate. A storage battery may even be considered as a very large capacitor; when fully charged, both sets of plates are good conductors and the electrolyte is a fair insulator.

Though it may be hopeless to attempt to list all possible types of capacitors, a nonexhaustive list is practical. Normally, fixed capacitors are classified according to the material used as the dielectric. Thus we find paper, plastic film, and ceramic capacitors in general use for coupling and bypass applications, with some ceramic units and micas employed in critical radio-frequency tuning tasks where air or vacuum is not necessary. Air and vacuum units are primarily used only in high-voltage radio-frequency tasks. Electrolytic capacitors are almost universally used in power supplies, where their relatively poor leakage characteristics can be tolerated; if the power supply voltage is too high for electrolytics, oil-filled units are used.

Let's take a more detailed look at the construction of these various general types. All, of course, consist of two plates and a dielectric, together with terminals or pigtails to permit connection to the plates, and usually including insulation to prevent either plate from shorting out to surrounding objects.

The oldest capacitor is the glass unit (Leyden jar). It is virtually unused today, however. Probably the most common capacitor is the ceramic disc type. Actually, ceramic is much too general a term, but it's the one in normal use. A ceramic material is anything glass-like, such as porcelain, china, and so forth. Some ceramics provide extremely high capacitance in small space. Others provide nearly perfect temperature compensation. Still others offer nondestructive breakdown when the voltage rating is exceeded. It all depends upon the exact ceramic used as a dielectric.

Ceramic capacitors come in two general shapes, known as **disc** and **tubular**. Tubular ceramic capacitors may be color-coded with the same code used for resistors, while disc ceramics are usually stamped with their value. In general, disc ceramics are used for bypassing signals from critical areas and for coupling between stages, while the more precise types of ceramics are furnished in tubular construction (although it's possible to find bypass tubulars, and precision discs, too). One major advantage of the disc construction over all other types of capacitor shape is the low lead inductance, which means that the capacitor will continue to behave as it should at frequencies so high that any other type of capacitor is acting like an inductor instead.

Small fixed mica capacitors perform many of the same tasks as precision tubular ceramics. These units consist of metal foil plates separated by thin sheets of the mineral mica. The entire unit is encased in molded plastic, and normally is about the size of a postage stamp.

At audio and power frequencies, the most common capacitor is the rolled tubular unit, which may use either waxed paper or plastic film as its dielectric. The paper units are made in both **conventional** foil and **metalized** types. The foil unit is composed of a sandwich of paper, foil, paper, and foil, wrapped into a tube (Fig. 3-13). When the paper dielectric is pierced by a voltage overload, it chars and short-circuits the unit. In a metalized capacitor, the metal film is deposited on

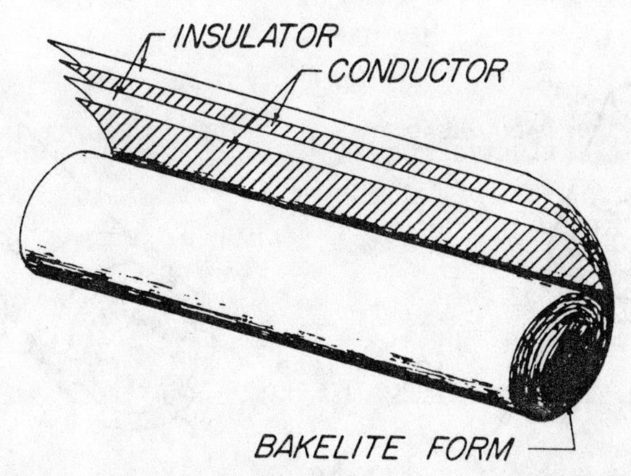

Fig. 3-13. A paper capacitor is so called because it uses paper as the dielectric. The foil type, shown here, consists of flat strips of metal foil interleaved with thin paper sheets (usually waxed). Paper capacitors usually range in value from several hundred picofarads to a few microfarads.

the paper dielectric and is extremely thin. A short circuit causes the metal film to burn away faster than the paper can char, preventing a short circuit and permitting the unit to remain in service. The resulting self-healing characteristic makes the metalized capacitor preferable.

Since all capacitor papers have minor defects such as pinholes, thin spots, and embedded conducting particles, the conventional unit must use two layers of paper rather than just one as its dielectric.

Characteristics of the plastic film dielectric are much more precise than those of paper, so that plastic film capacitors can be manufactured to closer tolerances and higher voltage ratings.

The electrolytic capacitor (Fig. 3-14) differs from other types in the fact that its dielectric is a chemical film formed on the surface of one plate. Because of this, electrolytic capacitors are all **polarized**; that is, they have positive and negative terminals, and will be destroyed or seriously damaged if hooked up the wrong way. Most electrolytics make use of aluminum for the plates or electrodes, and aluminum oxide as the dielectric.

For any given case size, the capacitance of an electrolytic unit can be increased by a factor of 2 to 4 times by etching the plate prior to its assembly. This results in what is known as an "etched cathode" capacitor. The increase in capacitance is

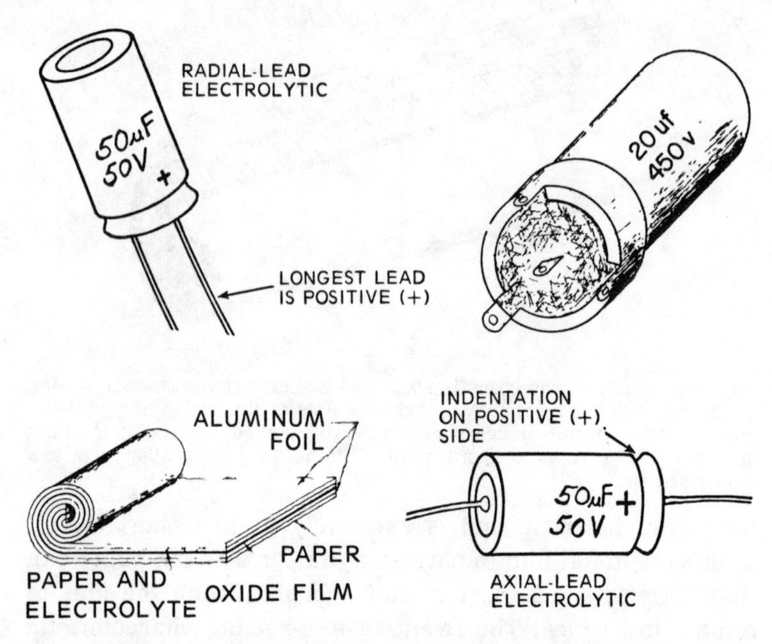

Fig. 3-14. The construction of an electrolytic might be quite similar to the paper capacitor. The positive plate consists of aluminum foil covered with an extremely thin film of oxide formed usually by an electrochemical process. This thin film acts as the dielectric. Next is placed a strip of paper impregnated with a pasty electrolyte, which serves as the negative electrode. Electrical contact to the negative electrode is provided by a second strip of aluminum foil placed against the electrolyte.

paid for by a corresponding increase in leakage and greater temperature sensitivity.

Use of electrolytics is confined exclusively to tasks in which the polarity requirement can be met and the low resistance poses no problem. Power supply filter action is one such application. Bypassing of low-value resistors in amplifier circuits is another.

Voltage Ratings and What They Mean

All capacitors are rated for voltage. Most carry two voltage ratings, one for **working** voltage and the other for **maximum** voltage.

The voltage rating of a capacitor indicates its intended working region. A capacitor rated at 10 volts (10V) is not intended to withstand application of 100V, and severe overvoltage almost always results in destruction of the capacitor. The manner of destruction depends upon the type. A paper capacitor may silently short out under excessive voltage, where an electrolytic more frequently detonates, spreading its electrolyte paste all over surrounding components.

The major factor determining final voltage ratings for any capacitor is the strength of its dielectric. A secondary factor is whether the dielectric is partially or fully self-healing or not. The metalized paper capacitor, for instance, is partially self-healing. An air capacitor, however, is fully self-healing. Overvoltage on an air capacitor causes a flashover between plates, but when the excessive voltage is removed, the capacitor is again ready for use.

Fully self-healing capacitors such as air and vacuum units usually have identical **working** and **maximum** voltage ratings, since no damage results from exceeding the voltage rating.

All other capacitors normally have different ratings for working voltage and maximum voltage. The maximum voltage indicates the value which cannot be exceeded without damage to the unit. The working voltage indicates the level at which the unit is intended to work. If only one voltage rating is given (as may be the case for some paper and plastic film units) it should be treated as a maximum rating, and the working voltage limited to something less to provide a safety margin.

The widest spread between maximum and working voltage is found in electrolytic capacitors. These units should be operated as close to the rated working voltage as possible, since because of their construction, they automatically adjust themselves downward. That is, a 450V capacitor operated for a long time at 300V becomes a 300V capacitor. This poses a problem, because it is still marked as a 450V unit; if 450V is then applied to it, the resulting bang as the overloaded unit detonates is an unpleasant shock.

Electrolytic capacitors are made in working voltage ratings up to 525V. If higher voltage ratings are required, they may be connected in series; but if this is done equalizing resistors must be used to insure that the voltage divides equally among the capacitors. This may make more sense after we examine simple circuits and voltage division in the next chapter.

INDUCTORS

In the previous chapter as we examined generation of alternating current, we saw that magnetism, motion, and electric force are always related so that with any two present, we get the third.

This implies that the movement of electric force (that is, the flow of current) must call forth a magnetic field; the electromagnet is based on this fact.

It also implies that any change in the flow of current must have a corresponding change in the associated magnetic field. This, in turn, implies that any change in the magnetic field must have an associated change in current. As it happens, these two implications join in the curious fact that any change in current flow opposes itself!

That is, the flow of current appears to have inertia. Just as an automobile or a fast-moving train cannot be stopped instantly, but must coast for some distance before halting, so does current tend to keep flowing once it is in motion.

This property is due to self-induced current, and is known as **inductance**. The unit of inductance is the **henry**, honoring the discoverer of resonance. While all circuit elements have inductance, since it's a property of current in motion, some are designed to maximize it, and others operate only because of it. Those which are designed to maximize it are known as inductors and commonly are called **chokes**; those which operate only because of it are known as transformers. We'll examine them separately; but first, let's see what inductance itself amounts to.

The conventional definition describes inductance in terms of the "line of magnetic flux." This is an imaginary line which early experimenters believed to have real existence. According to the usual definition, inductance is a measure of the number of such lines of flux which encircle the total current, per ampere of current. The henry is defined in this manner as being the number of "linkages" (lines of flux encircling total current) divided by current in hundred-millionths of an ampere.

The only problem with this definition is that the linkages, if they exist, do so in an unlimited region of space which may easily extend past the most distant galaxies, and what's more probably do not exist at all. Attempts to envision the counting of a specific number of such imaginary lines in an infinite space lead to mental corkscrews for most folk—but it's probably as good a definition as any, because inductance can be measured only by its effect in a circuit.

The true inductance of any component depends upon the material, shape, and size of the component, the amount of current flowing, the rapidity with which current flow attempts to change, the number, size, conductivity, shape, and proximity of all other conductors in the vicinity, and so forth. Thus, any measurement, calculation, or rating which claims to give "true" inductance is more than somewhat open to doubt.

The key point to remember about inductance is that it does impede any change of current flow in a circuit. Once the current flow becomes steady, however, the magnetic field becomes fixed and the inductance vanishes.

Because of this, inductance applies to direct-current circuits only for the brief intervals during which current flow is changing. In ac circuits, it is an important effect, because they never reach steady state. The following descriptions of chokes and transformers assume that only ac is involved, unless specifically noted otherwise.

Chokes

Chokes are used to provide a packaged inductance wherever it may be required. Normally they are used to "choke off" current flow in a circuit, and it is from this application that the name derives. Since inductance depends upon frequency, among other things, a choke is much more effective upon a high frequency signal than one of lower frequency, and tends to impede the high-frequency signal more. Like all forms of energy, electrical signals tend to take the path of least resistance, so the choke permits us to place opposition in one or more of several alternate paths and so steer signals where we want them to go.

Three major classes of choke or fixed inductor are normally available (Fig. 3-15). They are the filter choke, the

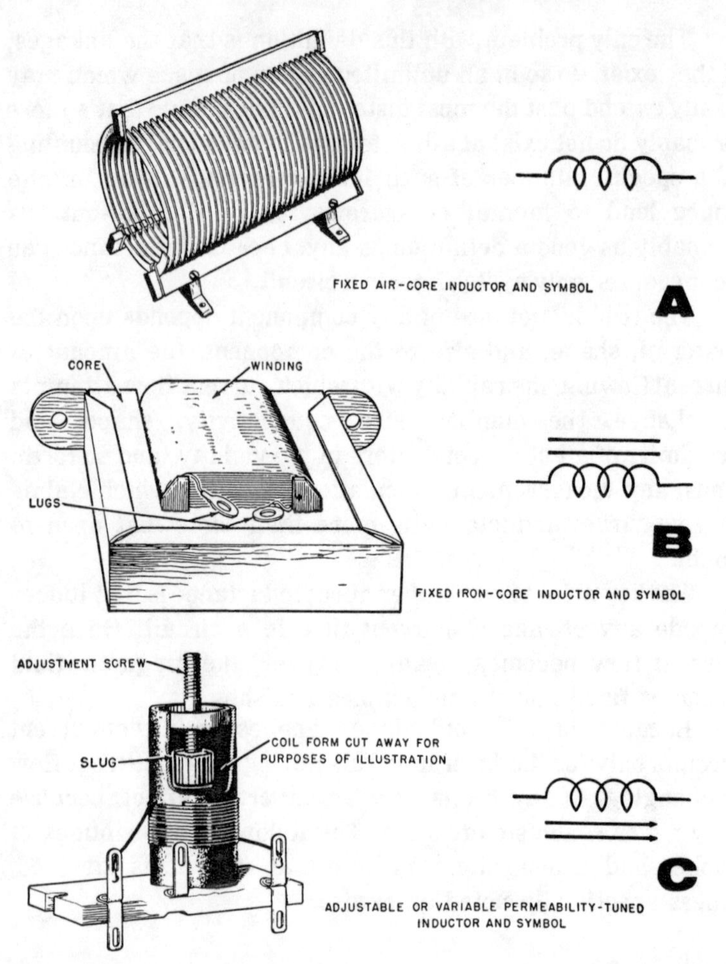

Fig. 3-15. Various inductors and their symbols. Like capacitors, inductors have a great many possible forms; the common element is a coiled length of wire.

audio-frequency inductor, and the radio-frequency inductor (often called an rf choke). These differ primarily in their intended frequency range. The filter choke is intended for use in power supply filters to separate dc from the ac ripple signal; it is the only one of the three designed to carry appreciable amounts of direct current in normal operation.

The audio-frequency inductor superficially resembles a filter choke in its construction, but is intended to provide packaged inductance for use in circuit coupling or tuning hookups.

The radio-frequency inductor or rf choke is similar to the audio-frequency inductor, with differences dictated by the different frequency range at which it is intended to be used. The audio-frequency inductor usually has an iron core; the radio-frequency version uses either air or ferrite in the core.

Since the action of an inductor is significant only in connection with other components in a circuit, we must postpone much of the explanation of these circuit elements until the next chapter.

Transformers

The transformer, invented by Faraday in 1831, has been called the world's most nearly perfect machine, since it can do its job at efficiencies that approach the theoretical maximum of 100 percent.

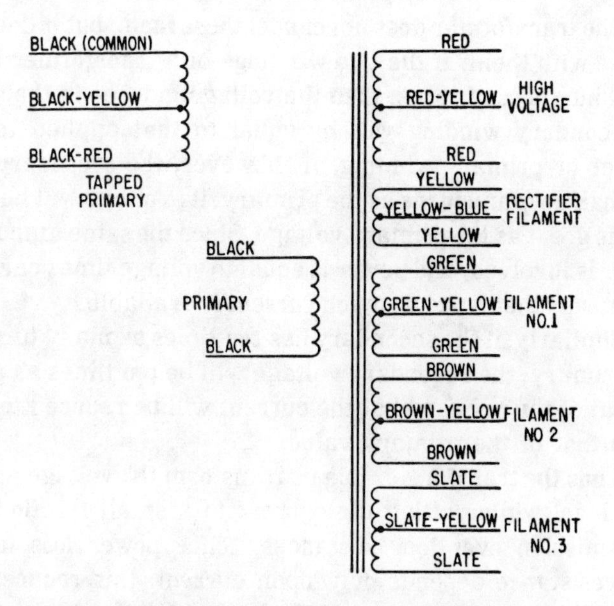

Fig. 3-16. Transformer consists of two or more inductors coupled so closely together that magnetic field of one causes induction of current in the others. If coupling is sufficiently tight, voltage ratio between windings is proportional to the ratio of turns in the windings. Winding to which current is applied is called the **primary**. If the primary of a transformer has 100 turns, one **secondary** has 500 turns, and another secondary has 10 turns, and 100V ac is applied to the primary, the first secondary will show 500V and the other will show 10V. (The traditional color-coding for small power transformers is shown.)

The basic principle of the transformer is simplicity itself; it consists of two or more windings, insulated each from the other, with a common core (Fig. 3-16). When current through one winding changes, the magnetic field in the core changes in step, and this induces current changes in the other winding. This makes it possible to transform an alternating voltage of one value into another value, while transferring this action into another—isolated—circuit.

The nationwide power distribution networks which furnish electricity to our homes and businesses could not exist without transformers. **Ohm's law** decrees that voltage drop in a resistance is directly proportional to the current flow through that resistance. As current increases, so does voltage drop. Since any power line has at least some resistance, this means that the more power we try to draw from the line, the more voltage is dropped by the line's resistance and so the less voltage is available to us at the load.

The transformer does not cancel these facts, but it does let us live with them. If the two windings of a transformer have equal numbers of turns, then the voltage induced in the slave or secondary winding will be equal to that applied to the master or primary winding. If, however, the secondary has only half as many turns as the primary, its voltage will be only half as great as the primary voltage. Since the same amount of power is involved, and power is equal to voltage times current, this means that twice as much current is available.

Similarly, if the secondary has ten times as many turns as the primary, the secondary voltage will be ten times as great as that of the primary but the current will be reduced to one-tenth that of the primary value.

Thus the transformer lets us transform the voltage up to a high level while cutting the current to a small fraction, for transmission over long distances. Since power loss in the line resistance depends only upon current, this reduces the losses to acceptable values. At the load end of the line, another transformer brings the voltage down and current back up. Most residences are served by a "pole pig" transformer which reduces voltage from 7200 to 240V. The 240V "source" is split and distributed as a pair of 120V lines with common ground, so that we may use "220-volt" air conditioners and heavy appliances, or "110-volt" bulbs and small appliances.

In addition to its power applications, the transformer is also used for coupling circuits. In all applications the quantity actually transformed is **impedance**, but we'll wait until the next chapter to delve deeply into just what this amounts to. Briefly, it's similar to resistance, and resistance is actually a special form of impedance. By transforming impedances, the transformer permits voltage or current levels to be set at will, and also permits coupling between devices which are, by themselves, incompatible with each other.

INDICATORS

In any practical applications of electronics, indicators are necessary to let the operator of the circuit know what is happening. This may include anything from the simple pilot bulb on a hi-fi amplifier, through the complex viewscreen displays of a radar installation.

Since we're examining only **passive** circuit elements right now, we cannot go into the many types of active indicators such as cathode-ray tubes. Among passive indicators, we find two major types—those which emit light, and those which move a needle. A third type, which emits sound, is seldom used. Among the light emitters, we have pilot bulbs, including incandescent lamps and neon glow lamps, and semiconductor devices called **light-emitting diodes**. We'll hold off discussion of LEDs until we deal with transistors, since much of the action is similar.

Pilot bulbs are available in incandescent and glow varieties. The incandescent type resembles a flashlight bulb or a miniature electric light, and that's just what it is—a metallic filament enclosed in an inert gas, which is heated white hot by current through it. Essentially, it's a resistor, and some circuits use a pilot bulb as a special temperature sensitive resistor, since the resistance changes with heat.

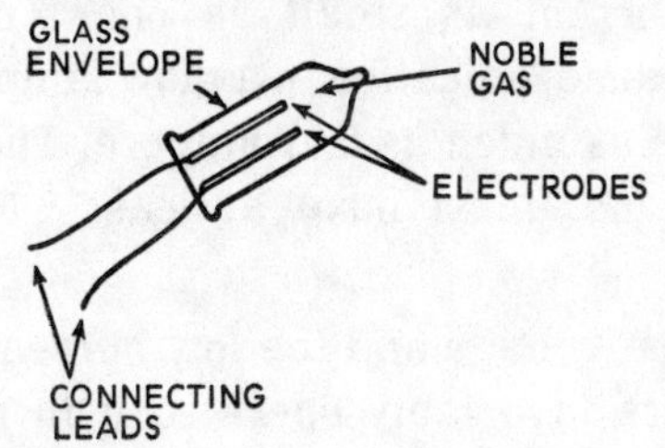

Fig. 3-17. Glow lamp consists of two electrodes sealed into glass tube together with noble gas, usually neon, at moderate pressure. At critical **breakdown** voltage, gas ionizes and permits current flow. Ionized gas around negative electrode then glows.

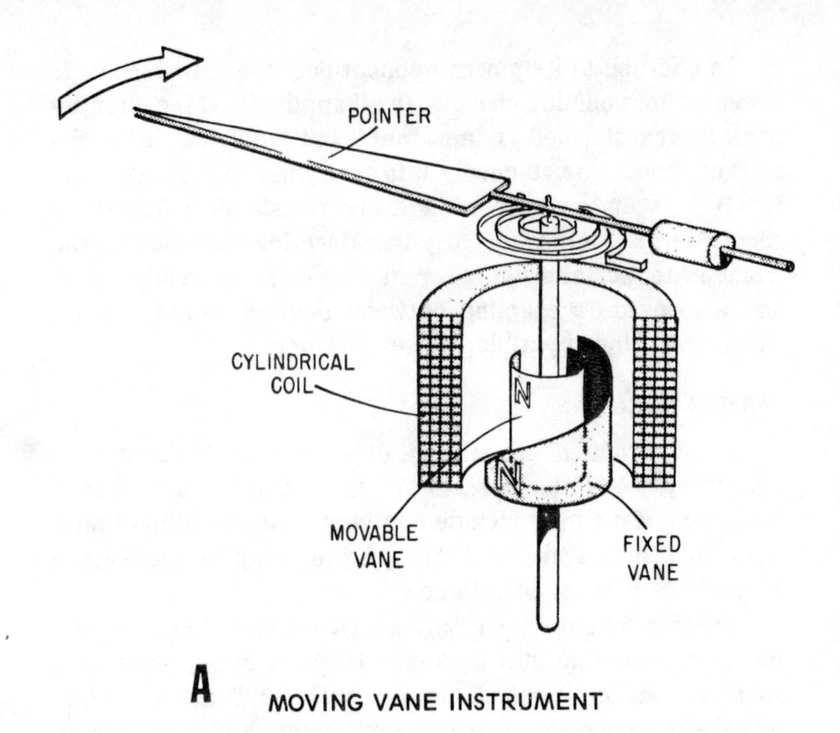

Fig. 3-18. Workings of two major types of meters are shown here. Moving-vane meter uses electromagnetic force to pull vane into coil. Needle attached to vane shows amount of movement and thus indicates current. Moving-coil meter uses magnetic repulsion between field of permanent

The glow lamp (Fig. 3-17) consists of two electrodes, insulated from each other, in a glass container filled with certain rare gases (usually neon) which break down or ionize at a specific voltage in the range from 65 to 200V. When the gas ionizes, the lamp glows. The glow originates in the gas around the negative electrode; if operated on ac, both electrodes appear to glow but actually at any one instant only one is active.

The great advantage of the glow lamp is that it operates at very low power levels. A current as small as one ten-thousandth of an ampere will cause indication in a glow lamp, and few if any of them require as much as 0.01 ampere. The incandescent, on the other hand, must have at least 0.06 ampere to function.

Because of the high ignition voltage and the low current levels required, glow lamps are invariably operated with a

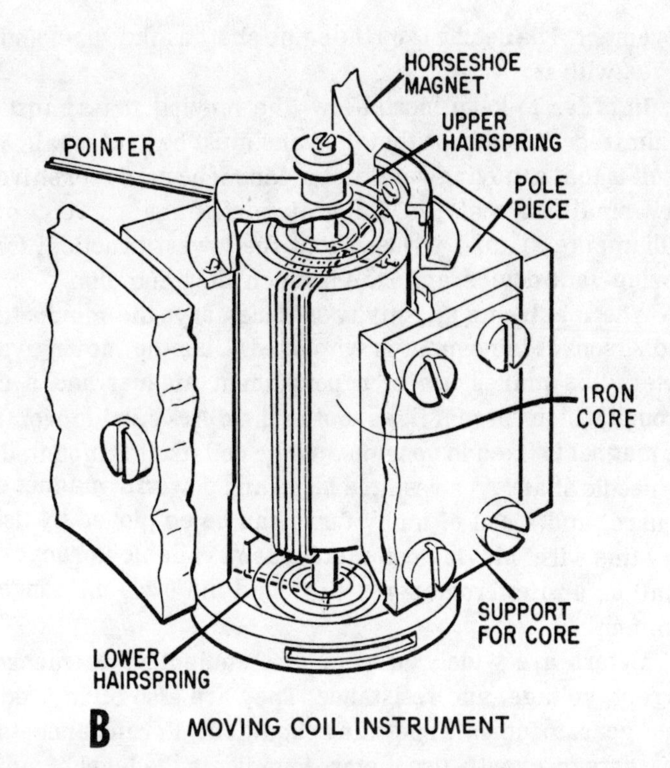

magnet and field accompanying current through coil. Needle is attached to coil. Of the two, moving vane is less costly but moving coil can be made more sensitive.

series current-limiting resistor. Some models have this resistor built into the bulb as a single unit. Others have it built into the socket. In most cases, though, it's up to the user to supply it separately.

The needle-moving indicators are usually called meters. They too come in two different major types. Both types (Fig. 3-18) depend upon interaction between magnetic and electric fields to convert electrical energy to mechanical motion, but they do it in different ways.

The least costly of the two is the moving-vane device, which uses a miniature magnetic vane mounted within a fixed coil of wire. Current through the coil causes the vane to move. Direction of movement depends upon polarity of the current, and the amount of movement is determined by the intensity of the current flow. The more current, the greater the movement. The needle is on the same shaft as the vane, and so moves with it.

In order to keep inertia low, the moving mass must be minimized. This means that the vane must be kept small. The result is that a moving-vane meter cannot be made sensitive to very small currents; a full-scale deflection current of 1 milliampere (1 mA) is just about the lowest practical for a moving-vane device, and 5 mA is much more common.

Where extreme sensitivity is necessary, the moving-coil or d'Arsonval movement is employed. Like the moving vane meter, this makes use of a permanent magnet and a coil through which current flows, but in the d'Arsonval movement the magnet is fixed in position and the coil itself is mounted on the needle shaft. As a result, a large and powerful magnet can be used, and a coil of many turns can be employed by using very fine wire. Moving-coil meters are available in ranges as small as 0-20 microamperes (uA) and the 0-200 uA range is common.

Meters are widely used in test equipment to measure current, voltage, and resistance. They are also often used in other gear, to indicate operating conditions. For instance, tape recorders frequently use meters to indicate the level of signal being recorded on the tape. Radio receivers include tuning meters to indicate accuracy of tuning. Automotive equipment such as tachometers and ignition analyzers use meters to indicate operating conditions of the engine. The list goes on and on; meters are among the most widely used passive elements in electronics.

Simple Circuits

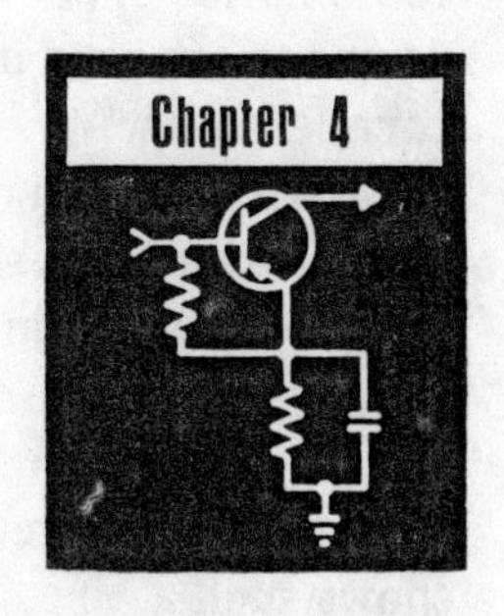

Now that we've made the acquaintance of the most common types of energy sources, and of most of the simple components which occur in electronic circuits of the passive variety, we are ready to move into practical applications and look at simple circuits.

The word circuit comes from **circle**, because a circuit (as we saw at the beginning of Chapter 3) is a complete path through which energy can flow. If the circuit is opened or broken at any point, energy flow stops and the circuit ceases to function.

We've met energy sources; they are those components of a circuit, such as a battery or a generator, which provide the electromagnetic energy for the circuit's operation.

Another term we will use more frequently as we proceed is **energy sink**. A sink is any circuit element which causes the electromagnetic energy to change to some other form. A resistor, for example, is a sink, because it changes the energy from its electrical form to a different form (in this case, heat).

In addition to sources and sinks, we have storage elements such as capacitors and inductors. These remove electromagnetic energy from the circuit for a time, but do not "lose" it by conversion of the energy to some other form. Instead, they store it, and release it back into the circuit later; thus, a storage element may act as a sink while it is collecting energy for storage, and as a source when it is releasing it.

Circuits may contain only one source and one sink, but circuits which are so extremely simple are rarely met in the practical world. Most circuits have a number of circuit elements.

Such multielement circuits can be of two types, known as series and parallel circuits. Actually, most circuits are a combination of series and parallel circuits, but can be broken down for analysis into strings of series elements, or banks of parallel units.

A series circuit is composed of elements connected end to end, like an old-fashioned string of Christmas tree bulbs, so that the current in the circuit must pass through first one element, then the next, and so forth all the way around the circuit. Any circuit in which all the circuit current passes through every circuit element is a series circuit. Figure 4-1 shows such a circuit, with one energy source and four sinks.

In a parallel circuit, each element is connected directly to the source, so that current through one element need flow through none of the others (and, in fact, cannot). For all practical purposes, a parallel circuit at this level is just the same as a number of super-simple single-element circuits which all share the same energy source. Current through each sink of a parallel circuit is independent of what happens in all the others, but full source voltage is applied to each element. Figure 4-2 shows several one-source, multisink parallel circuits.

The more modern type of Christmas-light string in which all other bulbs keep burning when one burns out is an example of a parallel circuit. In the older kind, with bulbs in series, failure of any bulb made the whole string go dark.

RESISTANCE AND CONDUCTANCE

Since resistors are one of the few circuit elements which always act as sinks (most of the rest may act as storage elements), we can start our examination of circuit action with them.

Resistance is the name we give to the property of transforming electromagnetic energy into heat. In a circuit, a resistor is used to either limit current to some desired value, or to establish a voltage level. In either case, the current and voltage interact. That is, if the resistor is being used to limit current, the voltage will vary as the current does, and the resistor must be of such a value that the current is limited to maximum value when the voltage is at its maximum. At any lower voltage level, less current will flow.

Conversely, if the resistor is being used to establish some certain voltage level, it can do so only so long as the intended value of current is flowing. If current becomes less, the

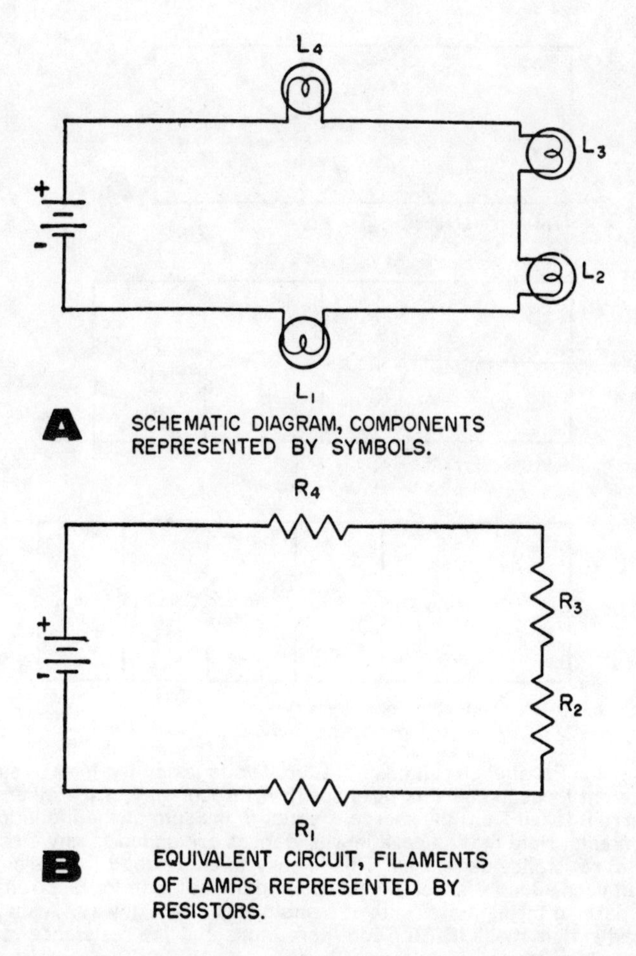

Fig. 4-1. In sketch A, the energy sinks are lamps; in B, they are resistors. In these series circuits, all the current from the battery (parallel lines with polarity symbols) is forced through every sink before it can be returned to the battery to complete the circuit. Voltage across any one sink is less than that across source, since each sink reduces it to some degree. If the filament of any lamp (A) breaks or if any resistor (B) becomes disconnected, the circuit is said to be "open," and no current can flow.

voltage level will rise, and if current increases, the voltage available at the load will drop.

The property we know as resistance is an inherent part of every circuit element, since it results from complex interaction between free and bound electrons within the atomic structure of all conductors. The device we know as a resistor simply concentrates this property in a measured quantity at a specific location in the circuit.

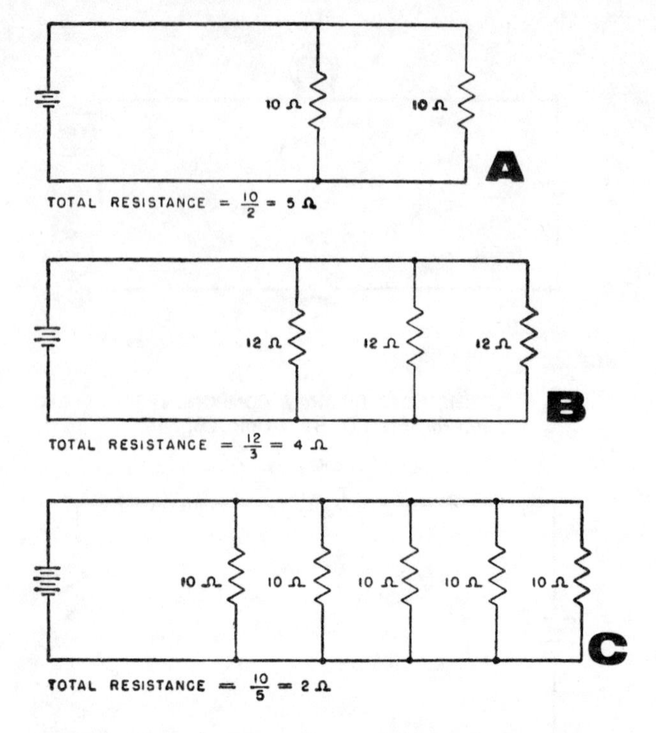

Fig. 4-2. Parallel circuit has full source voltage applied to each sink, and current through one sink does not flow through any of the others. Total current taken from the source is equal to the sum of the individual sink currents. Note that as parallel resistances are added to any circuit, the total resistance decreases. This is easy to understand if you remember that each added resistor provides an additional path for electrons to use in passing through a circuit. It's analogous to a highway: A single-lane road resists mass traffic; add more lanes and the resistance to traffic flow decreases.

The word resistance implies opposition to current flow. However, almost every word has a complement—that is, something that goes along with it but is the other side of the coin. For instance, if you limit current in a given circuit to 3.2 amperes, then flow of current above this limit is prohibited, but currents up to that level are allowed.

In the preceding sentence, the words **prohibited** and **allowed** are complements to each other. In one sense, they might be called opposites—and they have opposite meanings when they are referring to the same thing. For example, smoking is prohibited in gasoline storage areas, but allowed in many other places. The sense we are using here, however, is that prohibiting one thing automatically allows other things.

When used in this complementary sense, the ideas are called **inverses** of each other.

In electronics, every property which can be measured has an inverse. Frequently little use is made of these inverse properties, but sometimes they make things easier. Their existence came about primarily because mathematicians wrote the theories in the first place, and an inverse is as natural to math as a noninverse; both are artificial.

In the case of resistance, the inverse property is called **conductance**. The unit of resistance, the ohm, measures the degree to which the flow of current is opposed. The unit of conductance, the **mho** (which is, interestingly, **ohm** spelled backwards), measures the degree to which the flow of current is permitted. In most cases, the idea of resistance fits more naturally into observed and intuitive events than does conductance. However, we shall see shortly how conductance can make otherwise muddy ideas a little clearer.

Ohm's Law For Direct Current

As we have observed before, in any circuit the resistance, current, and voltage are all interrelated. The form of this relationship is known as **Ohm's law**, for the German experimenter who first expressed it.

Ohm's law states that the potential of the energy source in a circuit is proportional to the current through the circuit multiplied by the circuit resistance. When the units for measuring potential, current, and resistance are all properly chosen, it comes out to this statement: **voltage (E) equals current (I) in amperes times resistance (R) in ohms.** Stated algebraically, $E = IR$.

The great value of Ohm's law is that it allows us to determine any one of these three related quantities so long as we know the other two. Only minor algebra is needed to rearrange the formula to yield **resistance** if voltage and current are known ($R = E / I$), or **current** when both voltage and resistance are known ($I = E / R$).

Many memory aids have been propounded to help students remember Ohm's law. The one used by the author was simply the word "ear", which is what comes out if you try to pronounce the formula "E=I R" and consider the "=" to be silent. The other two versions are then derived from this one

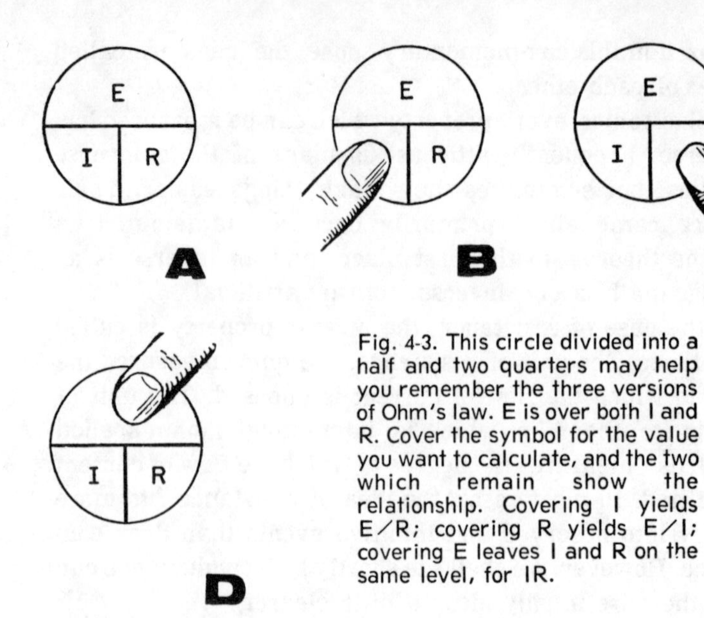

Fig. 4-3. This circle divided into a half and two quarters may help you remember the three versions of Ohm's law. E is over both I and R. Cover the symbol for the value you want to calculate, and the two which remain show the relationship. Covering I yields E/R; covering R yields E/I; covering E leaves I and R on the same level, for IR.

by algebra. Also, note that when laid out left to right, the letters are in alphabetical order—that makes this form of the equation simple to remember.

Another memory aid involves an eagle (E), an Indian (I), and a rabbit (R). The eagle flying above the desert sees both the Indian and the rabbit at the same level below him ($E = IR$). The Indian sees the eagle above the rabbit ($I = E/R$). The rabbit, similarly, sees the eagle above the Indian ($R = E/I$).

Figure 4-3 shows still another aid which uses a circle divided into three parts to provide the same illustration.

Ohm's law also gets into the definition of power, by an indirect route. Power is, by definition, the rate at which work is done, which in electronics comes out to be the **product of volts** and **amperes.** The "rate" is provided by current in amperes, since an ampere measures the rate at which charge moves. Work, in turn, is defined in physics as the movement of force. Potential in volts represents the force involved, and current introduces the movement.

However, it's not always convenient to deal with power in precisely this form. For instance, if you are attempting to determine how much power is dissipated in a 22-ohm resistor when half an ampere of current is flowing through it, you would first have to apply Ohm's law to determine the voltage

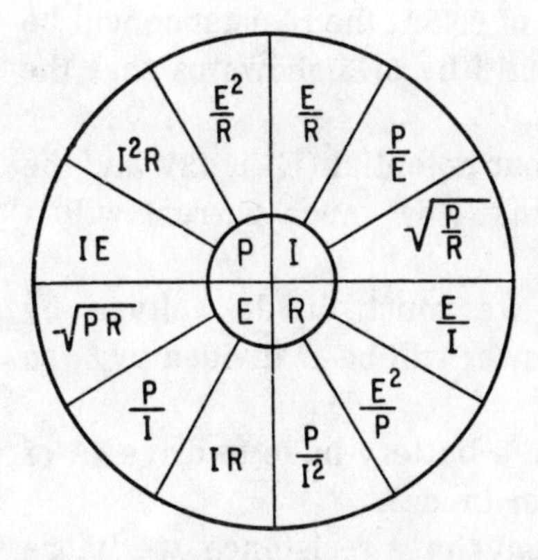

Fig. 4-4. Power (P) may be expressed in terms of alternate pairs of the three basic quantities of E, I, and R. In practice, you should be able to express any one of the three basic quantities, as well as P, in terms of any two of the others. The wheel here summarizes the twelve basic formulas you should learn. The four basic quantities are shown in the center. Note that there are three formulas for deriving any one of the four basic elements.

drop across the resistor ($E = I R$, so the voltage drop would be one-half (0.5) times 22, or 11V), then multiply voltage times current (0.5 x 11) to discover that the power involved was 5.5 watts.

By applying Ohm's law in advance, we add two more equations for power in terms of current and resistance, or voltage and resistance. That for current becomes $P = I^2 R$; for voltage, $P = E^2 / R$. The forms are illustrated graphically in Fig. 4-4. The first of these is almost as important to electronics as Ohm's law itself.

Ohm's law defines resistance as the ratio of voltage to current. What about the **inverse** of resistance, that commodity we call **conductance**? Surprisingly enough (to nonmathematicians), we get it by simply inverting the formula. G (the symbol for conductance) $= I / E$, or the ratio of current to voltage.

This means that conductance and resistance are **reciprocal** quantities, and their product for any specific case will always be exactly 1. In algebraic terms, $R = 1/G$, and $G = 1/R$. We'll need to remember this relationship before the end of this chapter, because it's necessary when we explore in depth the manner in which current flows in parallel circuits.

Now that we're full up to here with algebra, let's plug in some figures and see how Ohm's law works in one of the simplest of all practical circuits.

We have a 2-cell flashlight, and the bulb draws 450 mA (0.45 ampere). What is its resistance?

To work this out, we must know that each cell in the flashlight delivers approximately 1.5V, so the total voltage in the circuit is 3V. With current of 0.45A, the resistance will be E/I or 3/0.45 ohms. Dividing 3 by 0.45 shows us that the bulb's resistance is 6.67 ohms.

Let's try again. This time our potential (E) is 12V and the bulb's resistance (R) is 6 ohms. How much current will it draw?

For the current value, we must divide voltage by resistance. This means our answer will be 12 divided by 6; so the current will be 2A.

Finally, what voltage must a battery have to drive 4A of current through a 220-ohm resistance?

Since voltage equals current times resistance, we figure this one out by multiplying 4 times 220, and discover that an 880V battery is necessary. That comes under the heading of high voltage (and the resistor is a high-power situation, too, because it will be dissipating 4 times 880, or 3520 watts. That's somewhat more than the power required by the average electric heater).

Resistors in Series

When we connect two circuit elements in series, all the current which passes through one must pass through the other. We've noted this before. Ohm's law tells us that the current through a circuit depends upon both the voltage across the circuit and the resistance in the circuit. It also tells us that the resistance in a circuit depends on the voltage and the current.

Adding the second resistor to the single resistor of a supersimple circuit, to turn it into a series circuit, cannot increase the current through the circuit, because all the current which flows through the original must now flow through both. If anything, it will reduce the total current, because it must now flow through two resistors instead of one.

Similarly, changing the number of energy sinks in the circuit cannot affect the voltage of the source which feeds the circuit.

The only thing left which can be affected is the total circuit resistance. Since any current flowing in the circuit must now flow through both resistors, and the voltage has not been changed, the added resistance put into the circuit by addition of the second resistor must reduce the total current level. This reduction of current from the source then can be plugged back into Ohm's law to show that total circuit resistance increased.

The result is probably not amazing to anyone, since it only stands to reason that adding an extra resistor to the circuit would increase the circuit's total resistance. After all, when you add a resistive element to a series circuit, what you are actually doing is replacing a short length of wire (virtually zero resistance) with an element that will be an obstacle to current flow. If a whole string of resistors is connected as a series circuit, the total circuit resistance is the sum of the individual resistance values. When such a string is connected, as shown in Fig. 4-5, the voltage across the entire string is equal to the source voltage—which is another way of saying that it equals the current through the string, times the total resistance.

Across each individual resistor in the string, though, lower voltages will appear. In each case, the voltage across the individual resistor will be equal to the current through the circuit, times the value of that resistor. Thus the voltage drops across the individual resistors will total up to the full voltage of the source, which satisfies one of the primary laws of network theory. It's usually stated in textbooks as "the algebraic sum of the voltage drops around a closed circuit is zero," but what they're saying is that the sum of all the drops across energy **sinks** in a circuit is equal to the sum of the voltages of all **sources** in the same series circuit. The engineeringese version of the law assumes that when you switch your thinking from a sink to a source, you automatically change the sign of the voltage "drop."

The division of voltage across series resistors is called, naturally enough, **voltage-divider** action, and finds wide use in all phases of electronics. Voltage divider networks establish voltage levels in power supplies.

Even where voltage division is not necessarily wanted, it occurs whenever resistances are connected in series in complete circuits. One of the major problems faced by circuit designers is that of preventing unavoidable voltage-divider action from interfering with the action of their circuits.

A point to be watched out for when looking at practical voltage dividers is that the current drawn from the division

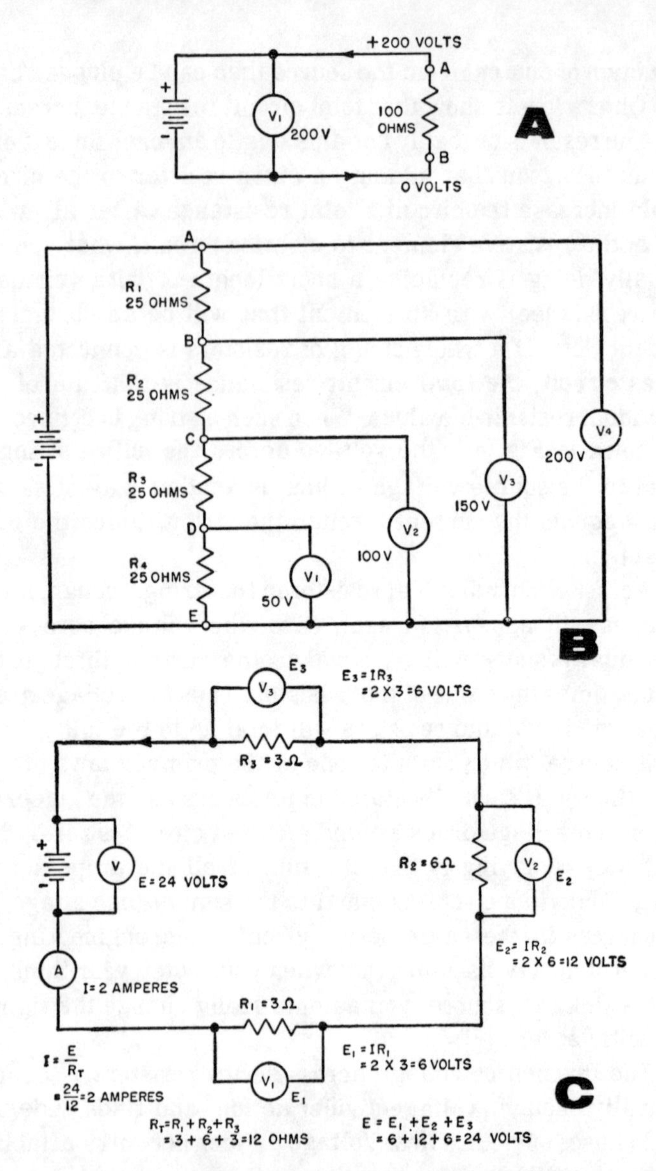

Fig. 4-5. In A, a 100-ohm resistor is placed across the 200V source, and the current through the resistor is 2A. Since the total resistance in B is the same as that of A, the total circuit current is the same (2A); but each resistor "sees" a different voltage with respect to ground. The voltage drop across each resistor in B is 50V, and the four 50V drops add up to the 200V provided by the source. In C, a 24V battery supplies three resistors whose values are not identical. The total current is 2A, but the voltage across each resistor depends on its value. The 6-ohm resistor drops 12V, and a meter would measure this amount; the 3-ohm resistors would each drop 6V. An important thing to remember: The sum of all the drops in a series circuit must equal the source voltage.

point contributes to the voltage drop across the upper leg of the divider, but not to that across the lower leg. In the simple voltage divider of Fig. 4-6, the value of the resistors is immaterial; so long as they are both the same, the total voltage is equally divided between the resistors. Let us assume the resistors are 100,000 ohms (100K) each, and explore a bit further. In this case, the total current through the resistors is 5 mA (0.005A), because current is equal to voltage (1000V) divided by resistance (200,000 ohms total). If you attempt to draw 5 mA from the tap point, the upper resistor is experiencing a 10 mA current drain which would drop the entire 1-kilovolt (1 kV) supply across it and pull the voltage at the tap point down to 0. If you draw only 1 mA from the tap point, this is still the same thing except not such an extreme case. Voltage at the tap will be less than 500V. We'll get into just how this happens later on in this chapter; for now, it's something to remember.

Resistors In Parallel

When we connect two circuit elements in parallel, none of the current which passes through one goes through the other, but the full supply voltage is applied to each. This means that the total current drawn from the source is greater than with either element connected alone. Ohm's law tells us that resistance depends upon voltage and current; with current increasing and voltage unchanged, total resistance must decrease.

For example, if we have two 12-ohm resistors and connect one across a 6V battery, Ohm's law ($I = E/R$) tells us it will draw 0.5A of current from the battery. When we connect the

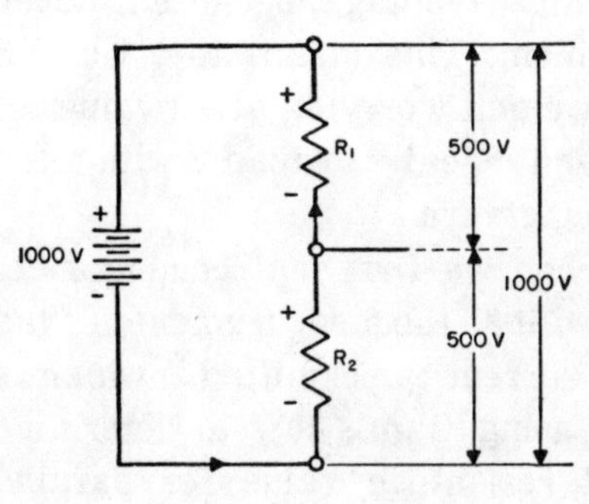

Fig. 4-6. When a pair of equal-value resistors are placed in series across a battery, the voltage will be evenly divided at the tap point. The value of the resistors will determine the total current the battery must supply; the lower the value of the resistors, the more current the battery will have to deliver. With a pair of 100K resistors, the total current will be 0.005A (total current equals total voltage divided by total resistance).

second one in parallel with the first, it draws another half-ampere, for a total current drain on the battery of 1A. Ohm's law, with these figures, shows that total circuit resistance of the two 12-ohm resistors in parallel is 6V divided by 1A, or 6 ohms.

The same relationship always holds true when two resistors of equal value are connected in parallel; the resulting total circuit resistance is half the value of each resistor. (Examine the simple parallel circuits of Fig. 4-2 again.) If three equal valued resistors are all wired in parallel, the resulting total is one-third the resistance of each.

It's not quite so straightforward if the resistors in parallel are not the same value, but the principles are the same. Let's retrace just what we did two paragraphs back: with known voltage (6V) and known resistance (12 ohms) we determined the current which would flow (0.5A). When we connected the second identical resistor, we added to this 0.5A the current taken by the second resistor, then went back to Ohm's law with known voltage (6) and known current (1) to determine total resistance (6 ohms).

Had the second resistor been 6 ohms rather than 12, we could have done the same things. Just as we did, we could find the current taken by the 12-ohm resistor as being 0.5A. We could then determine current through the added 6-ohm resistor, which is 6/6 or 1A. Adding the currents together gives us 1.5A total current, so the total resistance would be 6V/1.5A, or 4 ohms.

What we're doing with all this arithmetic is finding out how much current each resistor permits to flow in its own branch of the circuit, then adding the currents together. We don't really need to bother with the voltage, because it never changes throughout the problem. This means that we can simplify things quite a bit if we quit worrying about values of **resistance** temporarily, and deal instead with the corresponding values of **conductance**.

You'll remember that when we first met conductance, several pages back, we defined it as the measurement of "the degree to which the flow of current is permitted," which is exactly what we have been using Ohm's law to determine when figuring up the total resistance value for parallel resistors. What we must know is the degree to which each leg of the circuit permits current flow.

That is, if we use the conductance of each leg rather than the resistance, we can get at the total by simply adding together the values for all legs.

Note the similarity to the case of the series circuit. In the case of series resistors, the current was the same for each element but the voltage varied, and we obtained the total resistance values by summing up all the individual resistance values. In the case of parallel resistors, the voltage is the same for each element but the current varies, and we obtain the total conductance values by summing up all the individual conductance values.

This is our first meeting with the "complementary" concept of **duality**, which pervades all electronics. Every circuit, property, element, or concept has its complement, or **dual**, which is almost (but not quite) identical in operation. The differences are important, but so is the similarity.

Almost without exception, inverse quantities are duals, each for the other. Thus, conductance is the dual of resistance, current is the dual of voltage (because they are inversely related each to the other), and so forth.

Frequently, the way to understand a circuit which may be new to you is to look for its dual, which often turns out to be a circuit you already understand. This is exactly the way we are presenting the idea of parallel resistors, as the dual of series resistors, and we will continue this approach throughout the remainder of this book.

To get back to our parallel resistors, while it's much easier to grasp what's going on in the circuit by talking about adding up conductances, the practical fact remains that real resistors come marked with their resistance values; if you want to know conductance, you have to figure it out for yourself.

This isn't especially difficult, since for any specific resistance, the conductance (G) equals 1 / R, but it is arithmetic. The formula given in all the reference books for determining the value of a number of resistors R1, R2, R3,...R_n in parallel is actually a summation of their conductances, but they leave out the conversion to and from conductance and present it as follows:

$$\frac{1}{\frac{1}{R1} + \frac{1}{R2} + \frac{1}{R3 ... } \frac{1}{R_n}}$$

This is easy to write and may be easier to remember than the more explicit 3-step conversion, which first converts each resistance into a conductance by the formula $G = 1/R$, then determines total conductance by the equation:

$$G_T = G1 + G2 + G3 + ... + G_n$$

and finally converts G_T back to R_T by the formula $R = 1/G$.

The second way, however, since it spells it out step by step, may be easier to understand. Anyway, you have them both.

Examine Fig. 4-7 for a few minutes. As you can see, there is yet another method for calculating resistances in parallel circuits. The reciprocal method is shown for comparison.

Now that we have met the cases both of resistors in series and resistors in parallel, we are ready to look at the more practical cases, where resistors appear both in series and in parallel. One example of such a circuit might be the voltage divider we discussed a few pages back, where a load was being drawn from the divider tap so that the currents through the series resistors in the divider legs were not equal.

Look at the resistor-and-battery sketch of Fig. 4-8A. This may approximate what a resistive portion of an actual electronic circuit will look like. The resistors in sketch A can be rearranged to look like B, which bears a striking resemblance to the actual schematic diagram of the circuit (C). Practical electronic circuits do have all manner of series-parallel resistance combinations, and it often becomes necessary to calculate total resistance values in such cases.

The way to handle cases such as these is simple enough in principle, but often turns out to be downright tedious in practice. First each basic group or "circuit loop" which is either a pure-series or pure-parallel circuit is split off and converted to a single equivalent resistance. Then the resulting equivalent resistances are substituted back, and the whole process repeated. This goes on until you come up with a single total resistance value.

To show how it works, let's go through an analysis of Fig. 4-9 step by step. This is a moderately complex circuit—more

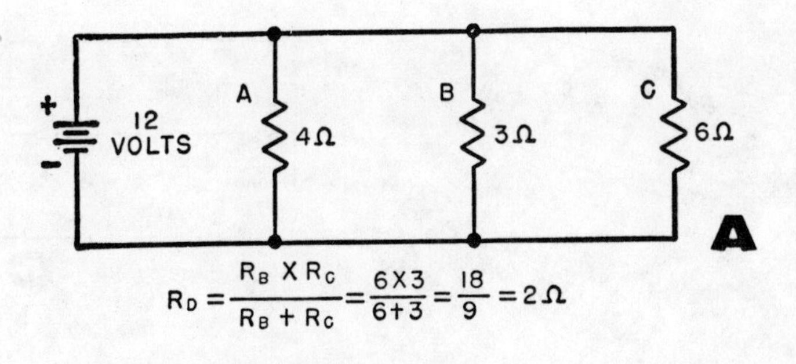

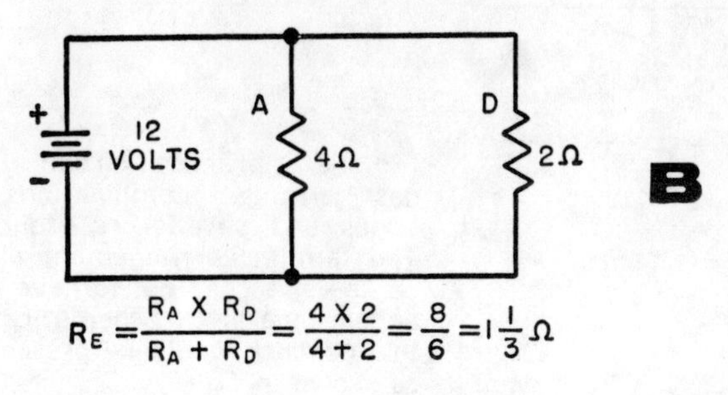

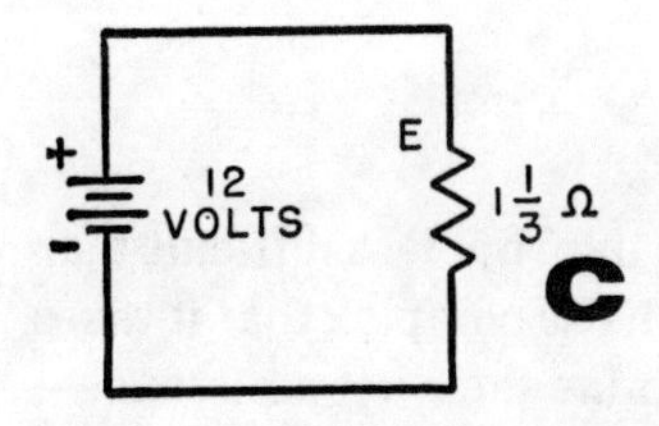

RECIPROCAL METHOD FOR COMBINING RESISTANCE IN THE ABOVE PROBLEM.

$$R_T = \frac{1}{\frac{1}{R_A} + \frac{1}{R_B} + \frac{1}{R_C}}$$

$$= \frac{1}{\frac{1}{4} + \frac{1}{3} + \frac{1}{6}} = 1\frac{1}{3}\Omega$$

Fig. 4-7. If you consider the circuit in A as a problem in which you must find the total circuit resistance, it is easy to see how you can simplify the problem by converting the circuit first to the circuit of B. The equation shown below A describes how the simplification process is done. B can similarly be converted to C, which is the solution to the problem. So another method of calculating total resistance from a pair of parallel resistances is to divide the product of the resistances by the sum of the resistances.

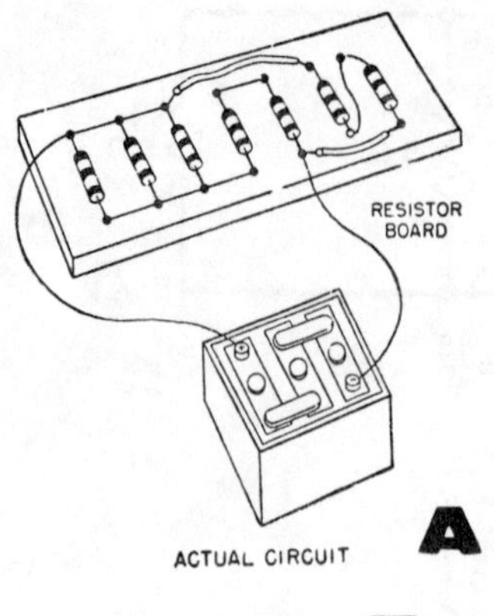

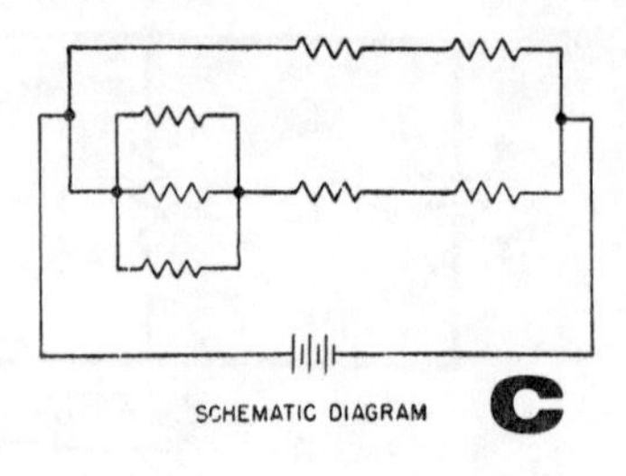

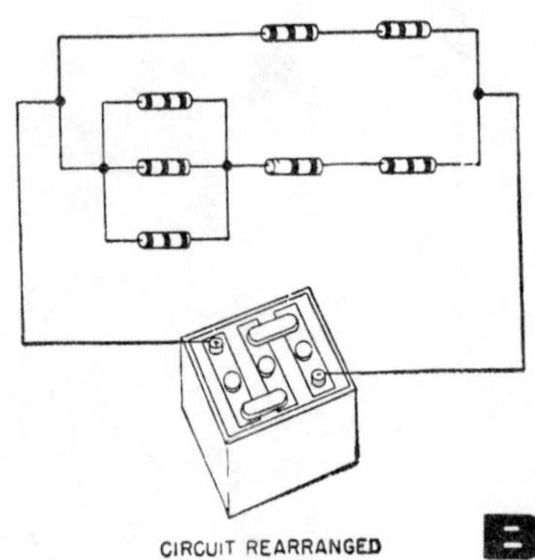

Fig. 4-8. A practical circuit often has "strange" combinations of series and parallel resistances. The familiar arrangement of A can be stretched out to have the appearance of B. Schematically, this network would be presented as shown in C.

complicated than many you will run into, but less difficult than others. It's possible to run into circuits so complex that it takes several days of calculation to determine their net resistance—but you're not likely to, until you get inside a design lab.

The simplest circuit loop in Fig. 4-9 is the parallel combination of R7 and R8. They net out to 12/7 ohm. We then add in the series value of R6, for a total resistance of R6, R7, and R8 of 54/7 ohm (conductance of 7/54 mho). R9, is parallel, has a conductance of 10/27 or 20/54 mho, so total conductance of the system of loops involving R6, R7, R8 and R9 is 27/54 or ½ mho, the same as 2 ohms resistance.

Since this entire loop system forms a single element in the series loop which also contains R4 and R5, the net resistance for R4 through R9 is 2 ohms (loop) plus 1 ohm (R4) plus 3 ohms (R5), or 6 ohms.

This 6-ohm net resistance (1/6 or 2/12 mho conductance) is in parallel with another combination made up of R2 and R3 in series. R2 and R3 together total 12 ohms or 1 / 12 mho, so that the conductance of all the loops taken together (so far) is 2 / 12 plus 1 / 12, or 3 / 12 mho. That's a net resistance of 4 ohms. This 4-ohm net resistance is in series with the 6 ohms of R1, so total net resistance of the circuit is 10 ohms. Current drawn from the battery is 2A.

We can work backward to check and verify all this, if we like. With 2A coming through the circuit, it must all flow through R1, giving us a 6x2 or 12V drop, and leaving 8V for the rest of the circuit. With 8V across the 12-ohm path formed by R2 and R3, the current through this leg would be 8 / 12 or 2 / 3 ampere, leaving 4 / 3 ampere for the rest of the resistors. Within the R2-R3 leg, with 2 / 3 ampere flowing, the drop across R2 would be 2 / 3 times 8 or 16 / 3 volts. That would leave 8 / 3 volts to drop across R3, which equals the 2 / 3 times 4 value given by Ohm's law, so all the energy in R1, R2, and R3 is accounted for.

Taking the 4 / 3 ampere at 8V through R4 gives us a 4 / 3 volt drop across the 1-ohm resistance of R4 and cuts the

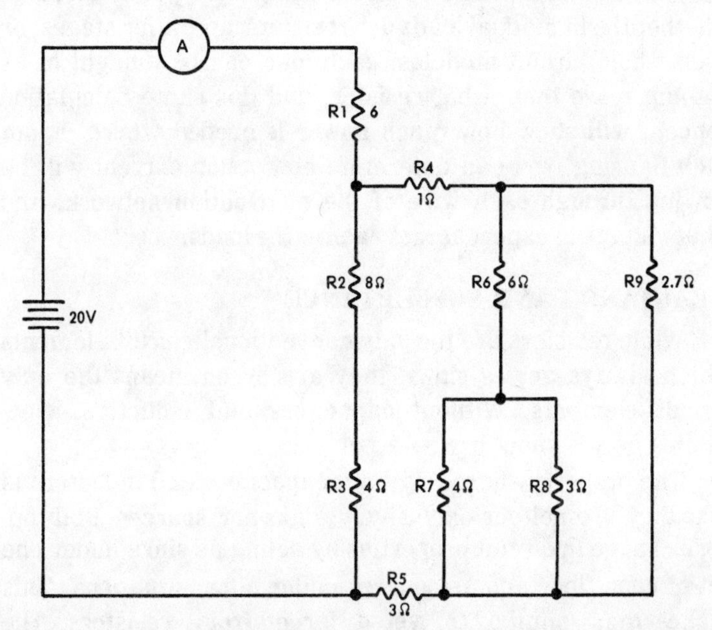

Fig. 4-9. This complicated network of series and parallel resistors draws the same current from the 20V source as would a single equivalent resistor. The value of the equivalent single resistor can be calculated using only series and parallel resistance techniques. Text shows how it's done; the principle permits analysis of any circuit.

voltage down to 20/3. The same 4/3 ampere through the 3 ohms of R5 takes away another 4V, leaving us 8/3 volt across R9. 8/3 volts divided by 27/10 ohms gives us a total current drain of 80/81 ampere through R9 and leaves us just a hair more than 1/3 ampere to go through R6, R7, and R8. That drops a fraction more than 2 volts through R6, so we have about 2/3 volt at 1/3 ampere for R7 and R8. Just a little less than 1/6 ampere goes through R7 (2/3 divided by 4), and a little over 1/6 ampere goes through R8 (2/3 divided by 3), adding up to the 1/3 ampere we had available and accounting for the whole 40 watts taken by the 9-resistor network.

If you work through this example several times, doing all the arithmetic, you should have a good grasp of the way in which circuits made up of both series and parallel resistances work to combine current drains and divide voltage across their various parts.

When you feel you have the system down pat, try working the two problems in Fig. 4-10. Solve for E, I, and R as stipulated in the drawings.

In most cases when we are dealing with power circuits, whether the individual loads are resistors, amplifier stages, or even whole circuit modules, each load can be thought of as nothing more than a big resistor, and this same calculation concept will show how much power is needed where. From such figuring, you can determine how much current will be flowing through each wire of the distribution network, and what voltage to expect across each of the loads.

REACTANCE AND SUSCEPTANCE

While resistors are the only conventional circuit elements which always act as sinks, they are by no means the only circuit elements. Without capacitors and inductors, electronics as we know it would not exist.

The property shared by both capacitors and inductors is that they are neither exclusively **sinks** nor **sources**. Both oppose change in circuit properties by acting as sinks under one set of conditions and as sources under other situations. This makes them similar to, yet different from, resistors. The distinction between the **opposition to change** (which characterizes both capacitors and inductors) and the **resistance to current flow** (of a resistor) is vital to this dif-

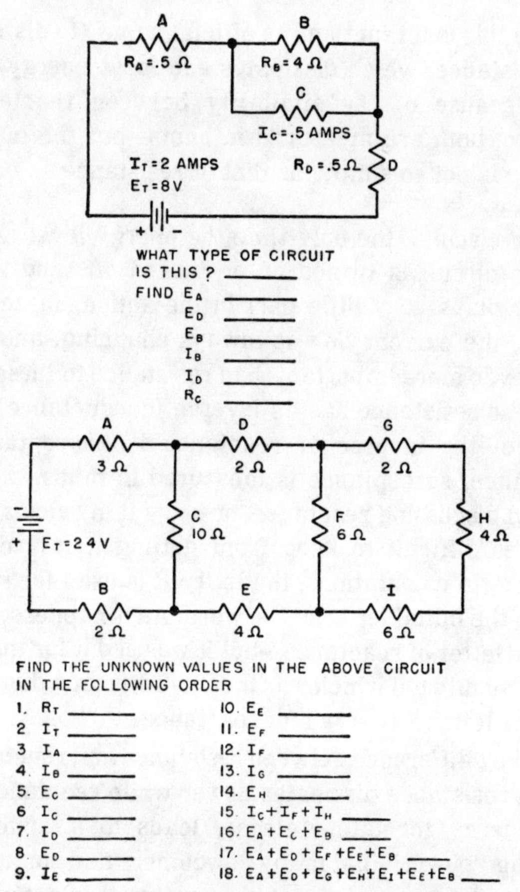

Fig. 4-10. Here are a couple of sample problems for you to experiment with. The initial letters in the symbols define the basic quantity (E, I, or R); the subscripts define the appropriate circuit element.

ference. Resistors dissipate energy by converting it into molecular friction which we know as heat. Once dissipated, the energy is gone forever from our circuit, and can return only after an extended cycle which takes it through the form of heat, to motion, through a generating plant, and back to electrical form. Capacitors and inductors, however, react to any change in current flow in such a manner as to oppose the change, and their reaction takes the form of temporarily storing energy if the change puts more energy into the circuit, or releasing energy from this temporary storage if the change is one which would reduce the amount of energy. The reaction, then, is one which stores excess energy and releases it later—in an attempt to maintain the energy level constant.

Since this is a reaction, its official name (to distinguish it from resistance, which dissipates and loses energy) is **reactance.** Because of the similarity between reactance and resistance, both are measured in ohms—but the measure of reactance is not so simple as that of resistance as defined by Ohm's law.

In dc circuits, the only time the energy level changes is when the circuit is turned on or turned off, and reactance normally plays very little part in the action. In ac circuits, however, the current flow is always changing, and so reactance is even more important than resistance to these circuits.

Just as resistance has its inverse (conductance), so does reactance—the inverse of reactance is **susceptance.** Like conductance, susceptance is measured in mhos.

When discussing reactance, or using it in calculations, it's sometimes difficult to keep from getting it mixed up with resistance. In calculations, the letter R is used for resistance since it's the initial letter of the word. As it happens, R is also the initial letter of reactance—but if we used it for this also we couldn't readily tell which was meant. To escape this problem, we use the letter X to stand for reactance.

The key difference between resistance and reactance—the fact that resistance dissipates power while reactance merely stores energy for later release, leads to a distinction in discussing "power" that many newcomers find confusing. In a dc circuit, the power dissipated in a circuit is determined by multiplying volts times amperes; the result, in watts, is the measure of power converted to heat.

When reactance enters the picture, however, not all the energy represented by the product of volts and amperes is dissipated. Some is stored in the reactance. Thus we have a new unit, the **volt-ampere,** in which to express wattless power. This is the product of volts and amperes when the only limiting factor in the circuit is pure reactance. When reactance and resistance are mixed (which is the usual case) some of the energy is dissipated and some is merely opposed. We distinguish, then, between **true** or **real** power, which is measured in watts, and **reactive** power, which is measured in **volt-amperes.**

We'll go into this in more detail a little later. First, we must meet **impedance,** which is the mixture of resistance and reactance. Right now, let's get a good grasp of what reactance amounts to and how we can handle it in circuits.

Delayed-Action Circuits

A major application of reactance in dc circuits is to provide a time delay between two events. This time delay is furnished by the reactance's opposition to the change of energy level in the circuit, which turns out to be constant for any specific combination of resistance and reactance.

To see how it works, let's take a resistor and a capacitor in series as shown in Fig. 4-11. If the resistor's value is zero ohms, the capacitor will theoretically reach full supply voltage instantly after power is applied to the circuit. In practice, of course, this cannot happen because any conductors which make up the circuit have at least **some** resistance. If the capacitor is fully discharged when the circuit is closed, it will at first appear to be a dead short across the power source and current through the circuit will be limited only by the resistance of the conductors. If the total resistance is only 0.01 ohm (a realistic value for typical workbench circuits) and the dc power supply is 12V, the current at the instant of connection would be 1200A, and power at this instant would be 14,400 watts, or 14.4 kW. Consider that even very modern homes are fused for a maximum of no more than 200A—and **that** only if they use electric power for cooking and heating. The capacitor is taking six times that current through a relatively small wire.

However, as soon as the capacitor accumulates even a little bit of charge, the applied voltage drops below the 12V

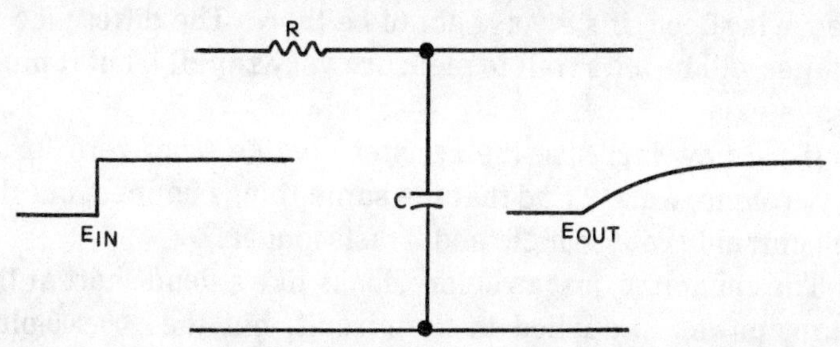

Fig. 4-11. When a sudden change in dc level occurs in a circuit having resistance and capacitance in series, the voltage across the capacitor cannot change instantly but requires time to reach its final value. This time delay depends upon the resistance and the capacitance. Because they determine the time, the product of R and C is known as the "time constant" of the circuit.

level of the source. When the capacitor voltage rises to 1V, the voltage that drives our current through the circuit drops to 11V (12V of the source minus the 1V on the capacitor). This cuts current to 1100A.

When the capacitor potential gets up to 2V, the voltage difference which drives the charging current drops to 10, and the current comes down to 1000A. As capacitor voltage climbs, current drops. When the capacitor gets up to 11.9V, the voltage difference is only 0.1V and the current is down to 10A. When the capacitor reaches 11.99V, the difference is 0.01 and the current is 1A. When capacitor voltage gets to 11.999V, current is down to 0.1A.

As the current drops, the time necessary to drive the same number of electrons into the capacitor increases, since current is merely the measure of charge moved per unit of time. Thus, each increase of voltage on the capacitor, which decreases the voltage difference, makes the next increase of voltage take longer to accomplish.

The initial current surge of 1200 amperes brings the capacitor up to almost full source voltage in an almost infinitesimally small fraction of a second. For instance, with 0.01 ohm of circuit resistance and a 100 uF capacitor, the voltage on the capacitor would be up to 7.56V just one millionth of a second (microsecond) after the circuit was closed, and would be at essentially full source voltage by the time five microseconds had elapsed. That's 0.000005 second to drop current flow down to an immeasurable trickle.

Yet despite this, the capacitor's voltage can never completely reach that of the source. No matter how thin the difference is sliced, it's always got to be there. The difference in voltages will be too small to measure very rapidly, but it must always exist.

If we now increase the resistor's value from zero up to 100,000 ohms, we will find that the same things happen, but the peak current is not so high, and it lasts longer.

The capacitor, just as before, looks like a dead short at the instant power is applied to the circuit, but the 100,000-ohm resistor limits the maximum current flow to 0.12 milliampere. With a current 10 million times smaller, it takes 10 million times as long to build up charge in the capacitor, so that 10 seconds after power is applied the capacitor voltage has only reached 7.56V, and 50 seconds are required for the difference between battery and capacitor voltages to be immeasurably small.

This 10-second time is completely independent of the voltage and current involved in the circuit; it is controlled entirely by the resistance and capacitance. Since it is constant even though voltage and current vary, it's known as the **time constant** of the circuit. In any circuit composed of a resistor and a capacitor in series, the time constant in seconds is defined as the product of the resistor's value in ohms and the capacitor's value in farads. At the end of one time constant, a charging capacitor will have reached 63 percent of its applied voltage. After one more time constant, it will have charged an additional amount equal to 63 percent of the difference. While no number of time constants is enough to bring it to full charge, the rule of thumb is that five time constants will bring it to within 1.0 percent of full charge; in practice, that's close enough.

In Fig. 4-12, curve A represents the **charging** time constant and curve B represents the **discharging** time constant. These curves are universally applicable to RC circuits. When discharging, in the first time constant, 63 percent of the voltage is discharged. In the next, 63 percent of what's left goes. In each succeeding time constant, 63 percent of the remainder leaks away. Since at the end of each time constant period, you still have 37 percent of what you started the period with, you'll never reach zero—but you'll get close.

Time constants work with inductance, too. The only difference between an RC time constant and an RL time constant

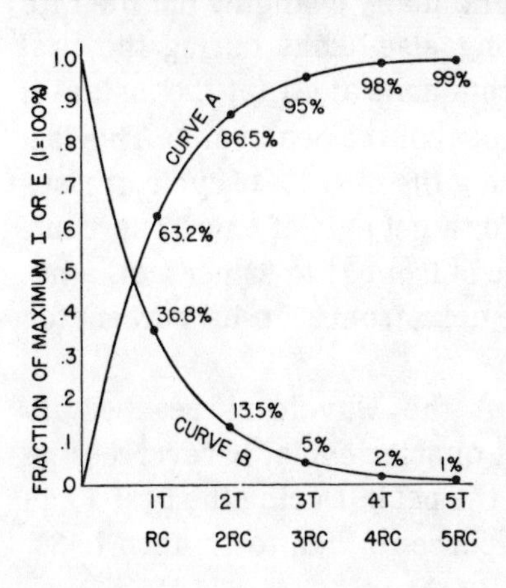

Fig. 4-12. Universal time constant chart. Curve A shows the varying rate at which a capacitor in an RC circuit charges. Theoretically, the charge never attains the full 100 percent source voltage, but after five time constants the capacitor charges to 99 percent of this level. Similarly, during discharge, the discharge rate slows as the charge leaks away; after five time constants, only 1 percent remains.

is that a charging RC time constant shows a slowly rising voltage into the capacitor and a slowly falling current through the resistor, while an RL circuit shows a slowly rising current into the inductor and a slowly falling voltage across the resistor.

If you deduce from this that capacitance and inductance are duals of each other, then you grasped the concept of duality perfectly when it was presented a few pages back. The effects of capacitance upon circuit voltage are exactly analogous to those of inductance upon circuit current, and so they are duals each to the other. This leads us directly to the two types of reactance, one associated with capacitors and the other with inductors, with capacitive reactance having its effect upon voltage, and inductive reactance affecting current.

Capacitive Reactance

Let's go back to the circuit of Fig. 4-11, with the resistor's value reduced to zero so that only the capacitor is effectively in the circuit, but replacing the battery with a source of ac, and see what difference this makes.

The first difference, of course, is that with ac instead of dc, the current is always changing rather than eventually reaching a steady state, and the instantaneous voltage is also in a state of continual change, so that the principles which governed the time-delay effects obtainable with dc are no longer in effect when ac is substituted.

Since the ac waveform follows the curve known as a sine wave, not only is the current continually changing but the rate at which it changes is changing also. Thus during the first 1/16 cycle, current changes from none at all (at the instant it reverses direction) up to 38 percent of the peak value. This is a net increase of 38 percent. During the next 1/16 cycle, it rises from 38 to 71 percent of peak, for a net gain of only 33 percent. In the next 1/16 cycle, the rise is from 71 to 92 percent, a net change of 21 percent, and in the next, from 92 to 100 percent for a net change of 8 percent.

At the quarter-cycle point the waveform reaches its positive peak. During the next quarter-cycle, it reverses the process by which it climbed to the peak. During the first 1/16 cycle it drops from peak to 92 percent, then to 71, then to 38, and at the half-cycle point crosses zero in the opposite direction.

If we sliced it into even smaller increments than the 1/16-cycle chunks we were looking at, the result would be the same. The change is continual, and so is the change of the rate of change.

Now let's see what this does to the charging of our capacitor. During the first 1/16 cycle the voltage increases from zero to 38V, and assuming that the time constant of the capacitor and circuit resistance is very much shorter than the 1/16 cycle time, the capacitor will be at essentially that voltage. During the next 1/16 cycle only 33V is added to the capacitor's charge, which means that a smaller number of electrons moved in the circuit—or that current was less. Similarly, during the third 1/16 cycle, when the voltage change was only 21V, the current was even less, and just before the voltage reached positive peak during the 1/16 cycle when the voltage change was only 8V, the current was nearly zero.

In fact, when the voltage passes through its peak and starts to drop, the current must pass through zero and begin to move in the opposite direction. There's no other way that the voltage can become lower, because charge has to leave the capacitor.

That is, when the voltage reaches its peak the capacitor is charged to essentially that same voltage. As the voltage of the source drops, the capacitor's voltage becomes greater than that of the source, driving current back into the source.

Note that with dc, the capacitor's voltage can never quite become equal to that of the source because the time constant requires that the difference keep being sliced ever thinner, and no matter how thin it's sliced, there's some left. With ac, when the voltage passes peak on each cycle, the role of **sink** and **source** swaps from the generator to the capacitor and back again.

The net result of all this is that the waveform of **voltage** in the circuit is not in phase with that of **current.** When the voltage waveform is beginning to drop from its positive peak, but is still positive, the current waveform is going through zero to the negative side of the balance. This is the same point which the voltage waveform will reach one-fourth cycle (90 degrees) later, so we say that the current in a capacitor leads the voltage by 90 degrees. It would be equally correct to say that the voltage across a capacitor lags 90 degrees behind the current through it, and this brings out that the capacitor's effect is to oppose (and thus delay) the change of voltage in the circuit.

This change of phase comes about because the capacitor is alternately storing and releasing energy. The amount of charge stored and released depends upon two factors: **applied voltage** and **capacitance**. Since current is just the measure of the amount of charge moved, these two factors will also influence the current in the circuit; but in addition, a third factor comes into play to help determine the total current. This factor is **frequency**, since that determines how many times each second the charge is moved.

Thus, the alternating current through a capacitor is determined by the applied voltage, the capacitance, and the frequency. If voltage is held fixed, the only factors left to establish a limit on current are the capacitance value and frequency, so these must be the factors which determine the reactance of the capacitor.

The relationship is formidable in appearance when you encounter the formula as it is usually presented, because of a conversion factor which pops up all over the place in physics. This factor is 6.28, or **2 pi**, which converts angular measurements from degrees to **radians**. Without going into detail about what a radian is (it's similar to degrees, in that it's a unit of measure for angles), we can account for the presence of this constant by the fact that ac is generated by coils going in circles. The frequency in hertz could be expressed in terms of revolutions per second just as well, and the quantity of **2 pi radians** is equal to one revolution. Thus, the "2 pi" factor converts from our conventional measurement of frequency in hertz, to a measure in radians per second.

The relationship between capacitance value, reactance, and frequency is constant. When each is expressed in the appropriate unit of measure, the product of the three is exactly 1. The units are radians per second for frequency, farads for capacitance value, and ohms for reactance. Since frequency is normally expressed in hertz, the conversion factor (2 pi) is put into the formula, and we are told that 2 pi times frequency (f) times capacitance (C) times capacitive reactance (X_C) equals 1, or as it is more usually expressed:

$$X_C = \frac{1}{2\pi fC}$$

In both expressions, f is frequency in hertz, C is capacitance in farads, and X_C is capacitive reactance, in ohms.

From this relationship we can deduce that at any one frequency, a large capacitor will have less reactance (offer less opposition to current flow) than a small one, and for any fixed capacitor value, reactance will be lower at higher frequencies. These facts make capacitors useful for separating low-frequency signals from those of higher frequency, since energy always takes the path of least opposition. By combining resistors and capacitive reactances, we can put together filters which will steer low-frequency energy in one direction and high-frequency current in another. This is done, for instance, in power supplies, where **decoupling networks** keep signals out of the common dc circuits by bypassing ac energy direct to ground through high-value capacitors.

Inductive Reactance

Since capacitance and inductance are duals, each of the other, it follows directly that everything we have learned about the effects of capacitors upon voltage in a circuit must apply also to the effects of inductors upon current. Just as a capacitor forces the current through it to lead the voltage across it by 90 degrees, the inductor forces the voltage across it to lead the current through it by 90 degrees (or, conversely, forces the current through it to lag the voltage across it by 90 degrees).

Like the capacitor, the inductor is alternately storing and releasing energy. Where the capacitor does its energy storage in an electric field, the inductor uses a magnetic field. That's the duality at work. Everything else, however, is the same.

Just as the reactance (X) of the capacitor is determined by capacitance and frequency, so is that of the inductor (L) fixed by inductance and frequency. However, the duality comes back into play, so that the formula becomes $X_L = 2\pi fL$.

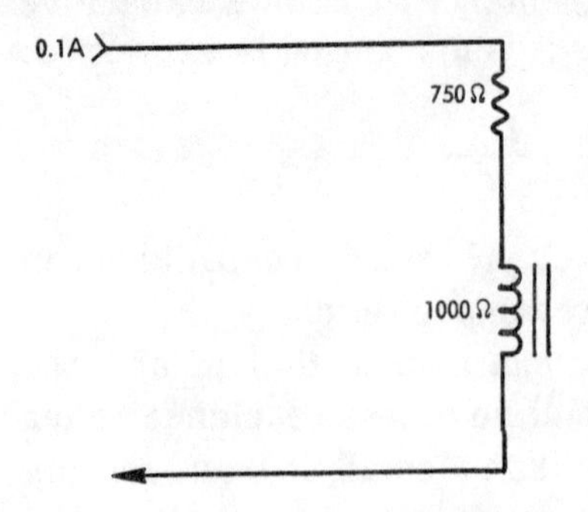

Fig. 4-13. When resistance and reactance are in series, the effective value depends on both, but isn't exactly obvious from either. The problem is that voltage across resistance and that across reactance are not in phase with each other, and so do not add up as you might expect them to. See text for details. While inductive reactance is shown here, capacitive reactance works the same way.

As either frequency or inductance increases, so does inductive reactance.

Inductive reactance can be used in filter circuits, just as we make use of capacitive reactance. However, it is seldom employed alone, because capacitors are less expensive for the same amount of circuit action, and also because capacitors make it easy to separate ac and dc while inductors do not.

The major application of inductive reactance is in conjunction with capacitive reactance. We'll be getting into this area a little later. First, let's examine what happens when we combine resistance and reactance, so that instead of either being pure, we have a mixture of the two. That's the only case we can ever run into in practice, because the **pure** concepts we have been studying exist only in theory. Actual circuits are a mix of a little bit of everything—including just enough mystery to keep us on our toes.

IMPEDANCE AND ADMITTANCE

While reactance and resistance are both measured in ohms, they're not the same thing—and so we cannot simply combine them if both happen to be present in a circuit at the same time. One of them represents dissipation of energy, while the other is a storage-release cycle in which the timing of the release acts to oppose current flow.

Figure 4-13 illustrates the problem with a resistor and an inductor in series. If the resistor's value is 750 ohms, and the inductive reactance is 1000 ohms, what is the net effect in the circuit?

A related question might be, how does the combination affect phase? The inductor causes current to lag voltage by 90 degrees, while in the resistor, current and voltage are in phase. What happens when these two effects are mixed?

On the face of things, it appears that the result must lie somewhere between the extremes. Thus the ohmic value of the combination should be something over 1000 but less than 1750, and phase shift should be between 0 and 90 degrees.

And while we're at it, what name should be given to this combination, since it's neither reactance nor resistance?

This final question is the simplest to answer. The word for the combination is **impedance**, since both resistance and reactance impede current flow. As we shall see, both pure resistance and pure reactance turn out to be special cases of impedance. Thus we can define impedance as **anything which is expressed as a ratio of voltage to current**, or (broadly) as anything expressible accurately in ohms.

Some texts use the term **resistive impedance** for resistance to emphasize that it's all impedance at bottom; they also use **reactive impedance** for reactance on the same grounds.

Sometimes it may be necessary to speak of resistive impedance when discussing a circuit which can be reactive, but not at the moment. This is the only sense in which we will use the phrase, however.

The complement of impedance is **admittance**; like conductance and susceptance, it is measured in mhos. Admittance can be thought of as **anything expressible as a ratio of current to voltage**, or simply **the inverse of impedance**.

Now that we have the words for it, let's see how resistance and reactance combine into impedance, using the circuit of Fig. 4-13 with 100 mA current flowing through it.

By Ohm's law we would expect the 100 mA (0.1A) through the 750-ohm resistor to give us a 75V drop, and it would be not only reasonable but exact to anticipate a 100V potential developed across the 1000-ohm reactance of the coil. However, this does not mean that we could measure 175V across the two in series, because the two voltages are not in phase with each other.

Since it's a series circuit, all the current which flows through one element must flow through the other, and this forces the currents through the resistor and the inductor to be in phase. In the inductor, however, the voltage leads the

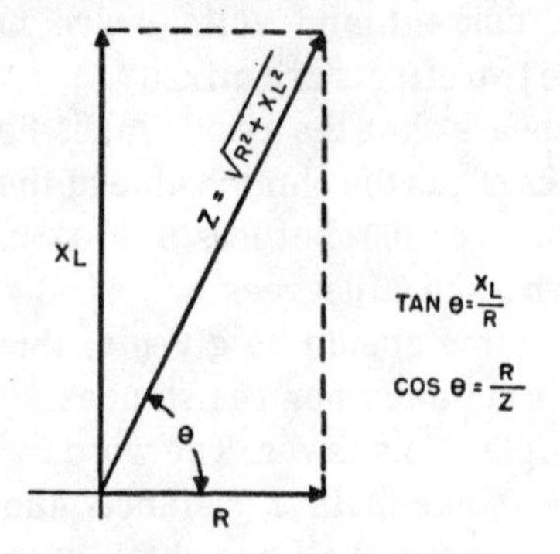

Fig. 4-14. Reactance-resistance chart shows how impedance can be calculated. In text case, value of impedance is 125 ohms and phase shift is 53 degrees. Simpler way of stating same impedance value is 750 + j1000 ohms, in which resistance and reactance values are independently stated. This not only bypasses most of the math, but offers other advantages as we shall see.

current by 90 degrees; in the resistor, voltage and current are in phase. Thus as the resistor voltage passes through zero going positive, the coil voltage is at its positive peak (let's call it 100 to avoid conversion problems, since we're really interested in how the voltages combine). Therefore, at this instant, the total voltage across the two in series is 0+100, or 100V.

A quarter-cycle later, when resistor voltage reaches its positive peak of 75V, the coil voltage is passing through zero on the downswing. Net voltage across the two series elements is 75+0 or 75V.

After another quarter-cycle, resistor voltage is back to zero going negative, and coil voltage is at its negative peak of —100. Net voltage of the pair is —100V.

We can see from this that both the phase and the amplitude of the net voltage are different from what we might have expected. One way of determining what they add up to would be to draw the two waveforms, one above the other, displaced by the 90-degree phase shift, and add them together at each point.

If we did, we would find that the total voltage across the series combination is 125, and that the phase shift is 53 degrees.

Another way to figure it out makes use of trigonometry, by plotting the resistance and reactive components of the pair at right angles to each other as shown in Fig. 4-14, then calculating the resultant distance which represents the impedance value, and the angle which represents phase shift.

The impedance value in ohms, as shown in the equation along Z axis of sketch, is equal to the square root of the sum of the squares of resistance and inductive reactance, and the phase angle θ is that angle whose tangent is equal to the ratio of inductive reactance to resistance.

If the reactance involved happens to be capacitive rather than inductive, the same rules apply, but the sign of the reactance is assumed to be negative rather than positive. This accounts for the duality of the two types of reactance.

There's another way to express impedance, which in many cases turns out to be simpler even though it's known as the **complex** expression (the name comes from its similarity to the "complex number" of mathematics). That's to express the resistive and reactive portions separately. Rather than saying that the impedance of the Fig. 4-13 circuit is 125 ohms at 53-degree phase shift, you could say that it was 750 + j1000 ohms. The first part, 750, expresses the resistive component, and the second part, +j1000, expresses the 1000 ohms inductive reactance. Had it been capacitive rather than inductive, the expression would have been 750 — j1000 ohms. Thus the complex expression provides all the details of how much resistance, how much reactance, and what kind of reactance.

Ohm's Law For AC

As we learned it earlier, Ohm's law applied only to dc circuits, because it dealt only with resistance and ignored the effects of reactance. However, we developed the concept of impedance by applying Ohm's law to reactance as well as to resistance.

Since impedance is the general case, which includes both resistance and reactance as special cases, it follows naturally that Ohm's law must work with impedance if it works with resistance. And, as it happens, that's the only change that must be made in order to make Ohm's law work for ac as well as for dc; everywhere that **resistance** appears in the dc version, **impedance** must be substituted for the ac case.

This is not necessarily quite as simple as it might appear at first glance, since the impedance of a circuit is determined in part by the ac frequency involved. If a number of different signals are present in the same circuit, each at a different frequency, the one circuit will have many different impedances, and the applications of Ohm's law will yield different results in each case.

While that might seem confusing, it's the basis of all filtering action, and what makes it possible to broadcast many different radio and TV programs, yet tune to any one while rejecting all the others. The key element to the tuning process is frequency, and what makes it so essential is that this property determines circuit impedance. We'll get into this in much more detail a little later.

Before we do, however, there's an important point about voltage and current measurement for ac that must be brought out. With dc, ambiguity is not possible. Voltage has only one value, and it doesn't change appreciably with time. Similarly, current is fixed (once the effects of time constants in the circuitry have stabilized).

With ac, however, the situation differs greatly. The value of either voltage or current depends entirely upon when you read it, since it's always changing and even reverses its polarity once every half-cycle. You can't very well take an average, either, for the average value of most ac signals is zero! (For every positive excursion, there's an equal-value negative excursion.)

One measure of either voltage or current is its "peak" value; that's the limit at which it changes direction. If a **peak** measurement is used, it's understood to refer to either negative peak or positive peak, but not both. Another measurement, similar but not identical, is known as **peak to peak**. This measures from negative peak to positive peak, and with normal ac waveforms is always twice the value of the peak measurement. For purposes of calculating resistance or impedance, it makes no difference whether peak or peak-to-peak measurements are used, so long as the same types of readings are used for both current and voltage, because what's important is the ratio of voltage to current.

When it gets to determining power in a circuit, though, difficulties arise. Assume a voltage of 2V and a current of 3A. The resulting power, calculated by the dc formula, would be 6 watts.

But if you measure two different circuits, using peak measurements in one and peak-to-peak measure in the other, the only way you could get the same number of units would be for the true voltages and currents to differ. Since peak-to-peak readings are twice as great as peak measurements, both the current and the voltage would have to be twice as great in the peak-measured circuit as in the peak-to-peak case, which would mean 4 times as much power. Yet the formula indicates 6 watts in each case.

To make it worse, if we measure power indirectly by determining how much heat is actually created by the power dissipation, we will find that neither the **peak 6 watts** nor the **peak-to-peak 6 watts** is actually as much power as a dc 6 watts, in terms of heat.

To keep ac measurements comparable to each other, the watt was made equal for both ac and dc, in terms of heat generation. For ac, this was called **effective** power because it produces the same heat effect as dc. Voltage and current standards were then developed for **effective** voltage and **effective** current, so that when multiplied they would produce effective watts directly.

The effective voltage in any ac circuit is equal to 71 percent of the peak voltage, or a little over 35 percent of the peak-to-peak value. It's also known as the root-mean-square or rms voltage, because it's determined mathematically by taking the square of each instantaneous voltage level, adding up all the squares, taking the average or mean, then taking its square root. The values just described are shown pictorially on the representative sine wave of Fig. 4-15.

Normally, ac voltage levels are expressed as rms values. Thus, the 110V household power, which is likely to be actually

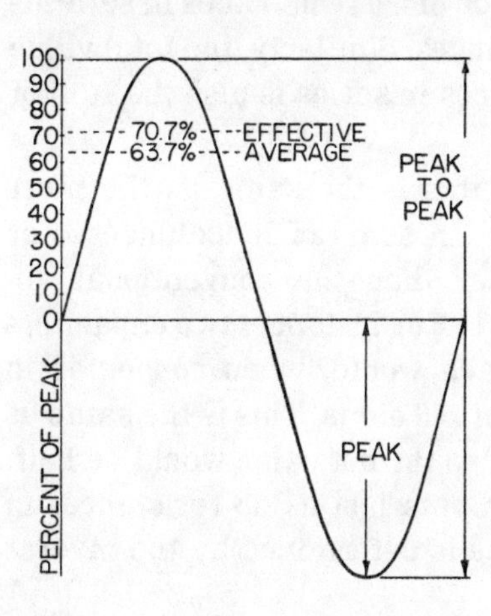

Fig. 4-15. There is always an ambiguity when an ac value is expressed, simply because an ac value is constantly changing. To standardize, we normally refer to an ac voltage value as peak, peak-to-peak, effective, or average. The effective value is based on the equivalent heat-producing ability of an available ac power when compared to a comparable dc source. Both **effective** and **average** ac values are calculated as a percentage of the peak value, as shown.

117V, is an rms value. Peak value of this everyday ac voltage is about 165 volts, and peak-to-peak value is around 330 volts.

When applying Ohm's law to ac, the rms values for voltage and current are normally used. However, if you happen to be interested in knowing the highest current value which can flow through the impedance during a cycle, you should use the peak values. Similarly, if you need the maximum voltage which can exist across an impedance for a cycle or more, you should use peak-to-peak values. It all boils down to using the appropriate set of values for the problem at hand.

Now that we have the problem of peak, peak-to-peak, and rms voltage and current values out of the way, let's see how reactances of opposite types combine in a single impedance value.

Resonance

Before we try to combine reactances of opposite types in a single circuit, let's see what happens when we combine reactances of the same type. That is, what is the result when we put two capacitors in parallel, or in series, or use inductors rather than capacitors?

While there are many ways to handle the problem, the simplest is to keep in mind that both resistance and reactance are special cases of impedance, and as a result any rules for combining resistances should also work for combining reactances.

Thus, the total value of two or more resistances in series is simply the sum of all the resistances. Similarly, the total value of two or more similar reactances in series is also the sum of all the reactances.

For inductive reactances, this is the same as the more conventional calculation based on sums of inductances. For capacitors, it's usually simpler than the conventional approach based on capacitor value. For instance, two capacitors having 100 ohms reactance each would, when connected in series, have a total reactance of 200 ohms. This is the same as one capacitor of half the value, so the net value would be half.

Reactances in parallel combine just as do resistances in parallel. That is, the total value is determined by the inverse property. The total susceptance of several parallel reactances is the total of all the individual susceptances.

If we're dealing with impedance, and it's expressed in the complex form, we can simply total up the separate parts for series circuits. That is, if an impedance of 50 + j30 ohms is connected in series with another of 30 + j100 ohms, the total value will be 80 + j130 ohms. If the impedance is not expressed in complex form, the problem becomes one beyond the scope of this work. Essentially, it's necessary to first convert to complex form, then do the combining, then convert back.

So far, we've been combining reactances only if they happen to be of the same type. What happens if we have the RLC series circuit shown in Fig. 4-16? In the circuit shown, assume the source frequency (circled sine wave) to be 1000 Hz; assume the resistor value to be 10 ohms, the inductor reactance to be +j200 ohms, and the capacitor reactance to be –j300 ohms.

Since it's a series circuit, we could reasonably expect the total reactance to be the sum of both the +j200 ohms of the inductor and the –j300 ohms of the capacitor. Keeping the signs in mind, the resulting total would be –j100 ohms. That would indicate that the capacitor's reactance had completely canceled the effect of the inductor, and still had some capacitive reactance left over, for a circuit impedance of 10 – j100 ohms.

As it happens, that's just the way reactances of opposite type combine. And since inductive reactance increases with frequency, while capacitive reactance decreases, for any LC combination there must be at least one frequency at which the reactances cancel completely.

If in Fig. 4-16, we increase the frequency from 1000 Hz to 1500 Hz, the inductor's reactance climbs to +j300 ohms and the capacitor's reactance drops to –j200 ohms. Circuit impedance the capacitor's reactance drops to –j200 ohms. Circuit impedance at this frequency is 10 + j100 ohms. Clearly, the point of balance for this circuit is somewhere between 1000 and 1500 Hz.

At a frequency of 1250 Hz, the reactances are +j250 and –j250 ohms respectively. Thus the circuit impedance at this frequency is 10 + j0 ohms. That is, all the reactance is canceled out, and the resulting impedance is purely resistive.

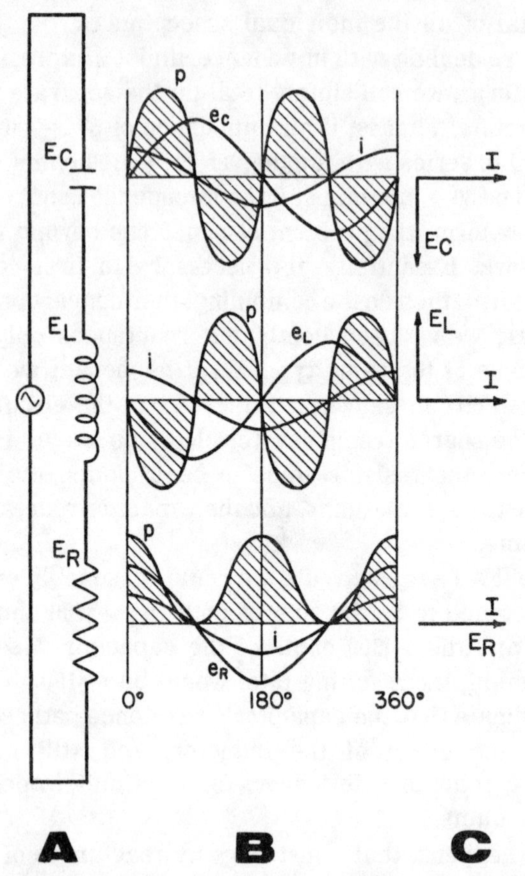

Fig. 4-16. Circuit supplied from 1000 Hz source containing R (10 ohms), L (j200 ohms), and C (—j300 ohms) has net impedance which depends upon all three. Reactances cancel, so that effective reactance in this circuit is —j100 ohms, with resulting net impedance of 10 — j100 ohms for circuit taken as a whole. At frequencies other than 1000 Hz, reactances differ, and so circuit impedance varies. This makes tuning possible.

The importance of this effect is more clearly seen when the circuit impedance is expressed in the noncomplex form. It happens that when reactance and resistance bear a 10-to-1 or higher ratio to each other, the impedance is so close to being identical to the larger value that for all practical purposes, the smaller component can be ignored.

If we do so, we find that at a frequency of 1000 Hz, the circuit seems to be a capacitor with 100 ohms reactance. At 1250 Hz, it appears to be a 10-ohm resistor. At 1500 Hz, it seems to be an inductor with 100 ohms reactance.

If we apply three different ac signals, one at each of these three frequencies, to the circuit simultaneously, we will find that most of the current in the circuit is at 1250 Hz. The 100-ohm reactances tend to limit current at the other frequencies, but at the frequency of balance (known as the series-resonant frequency or just **resonance**) the only limiting factor is the resistance.

If we leave the capacitor and inductor values the same but reduce the resistance to 1 ohm, the effect is even more marked. The reduction of resistance has no detectable effect on the 1000 and 1500 Hz signals, which were limited by the reactances. At 1250 Hz, though, current increases tenfold.

Thus a series-resonant circuit can separate signals at the frequency for which it is resonant. This frequency can be changed by adjusting values of either the capacitor or the inductor (or both). The effectiveness with which the separation is performed depends upon the ratio of reactance to resistance. Since reactance is canceled out in the complete circuit, this **quality** factor or **Q** is defined in terms of inductive reactance only; it's the ratio $X_L:R$. While the **Q** stands for quality factor and the ratio measurement was its first use, you'll also find this factor appearing all over electronics, just as we found that **ohms** really refers to the ratio of voltage to current. **Q** is the ratio of energy stored to energy released, per signal cycle, and in the case of a resonant circuit, the ratio of reactance to resistance gives us the appropriate value.

We have seen how reactances of opposite signs cancel each other at the resonant frequency when connected in series. What happens if they are connected, instead, in parallel?

It's the same story as any other parallel circuit. Instead of the reactances canceling, it's the susceptances—the inverse properties—which cancel. The result, as in the series circuit, is pure resistance, but unlike the series circuit, the resistance is not lower than the off-resonance impedance, but is higher.

The reasons for this are somewhat obscured by all the rigmarole of measurement, but become clear when the nature of reactances in general is reconsidered. Remember that the big difference between reactance and resistance is that while resistance dissipates energy, reactance stores it and returns it to the circuit a little later, out of phase.

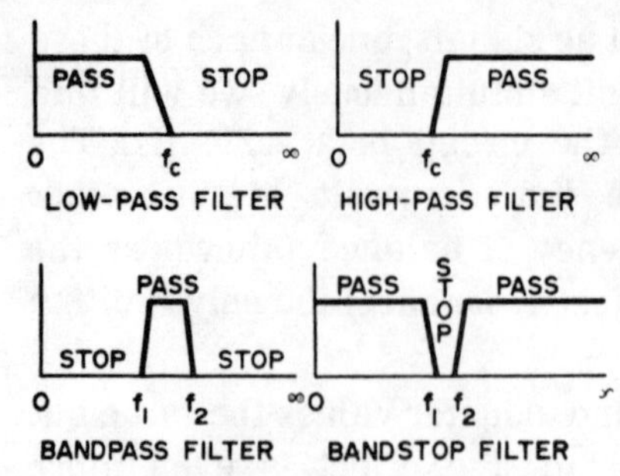

Fig. 4-17. Idealized frequency-response charts illustrate principal characterisitcs of low-pass, high-pass, bandpass, and bandstop filters. Each separates applied input signals into groups which are either passed or stopped. Frequencies which mark boundaries between these bands are known as cutoff frequencies.

When opposite reactances are connected in parallel, each of them is storing up its own kind of energy at its own time, and releasing it a little later, out of phase. However, just as each is releasing it, the other is ready to accept it, so very little energy gets out of the circuit—or is accepted from any external circuits. Each reactance's stored energy is almost adequate to supply that required by the other. All that must come from outside is the relatively small difference.

The result is that in the case of a parallel resonant circuit, the **Q** factor is an impedance multiplier. If either reactance would be, say, 1000 ohms at the resonant frequency, and the circuit **Q** is 100, then the effective impedance of the parallel resonant circuit is 100 times 1000, or 100,000 ohms.

Filters

We mentioned earlier that the multiple impedances presented by the same circuit to signals of differing frequency was the basis of all filtering action and tuning.

Tuning is usually accomplished by adjusting a resonant circuit of the type discussed in the preceding section. The subject of filters, though, is much broader.

In general, a filter is a special type of circuit made up of several reactances. The purpose of a filter is to separate signals of differing frequencies. Usually, one group of signals is permitted to pass through the filter, and all other frequencies are bypassed to ground.

Depending upon the group or **band** of frequencies to be passed, filters are classified as low-pass, high-pass, bandpass, and bandstop (Fig. 4-17). A low-pass filter passes all frequencies from dc (or very low frequency in some cases) up to its cutoff frequency. A high-pass filter passes frequencies above its cutoff frequency. A bandpass filter may be considered as a mating of the high-pass and low-pass designs so that signals which are above the lower cutoff, yet below the upper cutoff, are passed. A resonant circuit is a crude form of bandpass filter, but the term is usually reserved for more complex forms in which the passband (band of frequencies passed by the filter) is very wide in relation to the center frequency.

We can see how filters work by examining operation of a low-pass unit, followed by a look at a high-pass circuit. The low-pass filter is typically found in dc power supplies which convert ac household current to dc for powering other circuits. The high-pass filter is frequently encountered in interference filters for TV reception, designed to pass TV signal frequencies while rejecting lower-frequency interfering signals from CB and ham radio activities.

Figure 4-18 shows a typical **unbalanced** low-pass filter composed of two capacitors and one inductor. At very low frequencies, such as dc, reactance of the capacitors is very high and that of the inductor is very low, allowing signals to pass straight through the inductance virtually unimpeded. At high frequencies, reactance of the capacitors is low and that of the inductor is high, so signals get short-circuited without being passed. By voltage-divider action, the output at the cutoff frequency is just half that at the low frequency.

In all filters, the cutoff is not a sharp action. It's a gradual thing. With two reactances involved, the power passed will drop to half with each successive doubling of frequency. With more reactances, the drop can be made steeper. In practice, only enough reactance is supplied to provide the required performance; no attempt is made to reach an ideal condition.

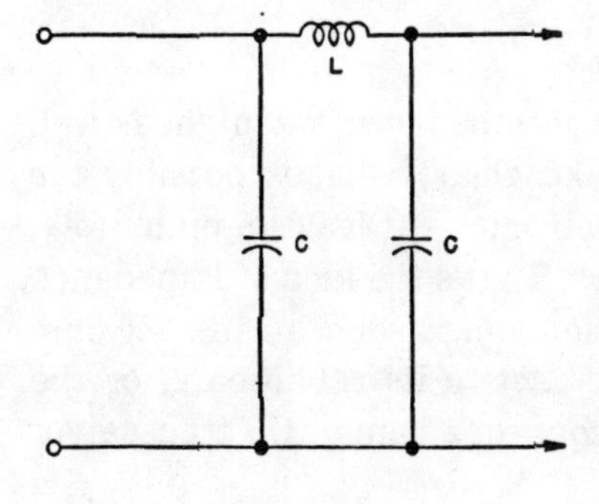

Fig. 4-18. Basis of low-pass filter is fact that impedance of inductor is high at high frequency, while that of capacitors is low. Filter circuit puts capacitors in parallel with input and output terminals, and inductor in series. At low frequencies filter is effectively not there because series reactance is low and parallel reactance is high. At higher frequencies, however, situation reverses.

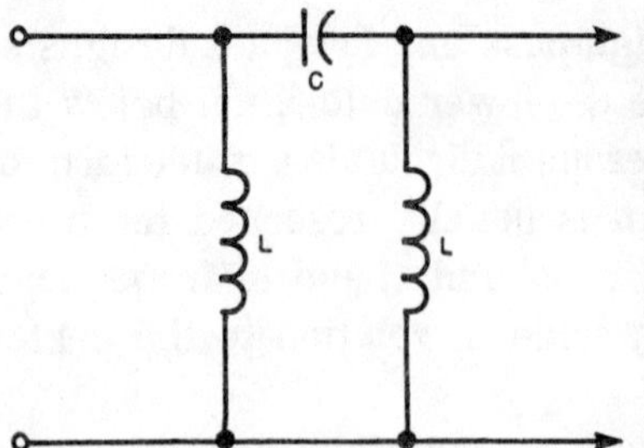

Fig. 4-19. High-pass filter action is just the opposite of low-pass filters. Network offers short circuit across both input and output together with open circuit through the filter for low-frequency signals. When signal is above cutoff frequency, situation reverses, as impedance of inductors rises and that of capacitor falls.

The high-pass filter shown in Fig. 4-19 follows the same principles to achieve reversed results. At low frequencies, the low reactance of the inductors shorts out the signal while the high reactance of the capacitor is effectively an open circuit. At high frequencies the situation reverses. In between, at cutoff frequency, the output is still just half the voltage of the input.

The inductors in these filters can be replaced by resistors, but the cutoff rates will be even more gradual.

Bandpass filters essentially combine a low-pass filter having a cutoff equal to the **upper** bandpass limit desired, and a high-pass filter with cutoff frequency equal to the **lower** bandpass limit. The complementary network is the bandstop (band-rejection) filter. These "duals" are shown in Fig. 4-20.

The subject of filter design is actually much more complicated than this discussion would indicate, but the complications come about because a designer has to meet many requirements in addition to that of cutoff frequency. He must be certain that his filter will work when put into a larger circuit without undesired interaction between its reactances and others, and so forth. The principles, though, are no more complex than they appear to be here.

Transmission Line Impedance

While we're on the subject of impedance, we might as well bring up (and attempt to make clear) what is possibly the most confusing concept of electronics—at least to many folk, not all of whom are newcomers. That's the idea of impedance as it applies to rf transmission lines, such as the 300-ohm twinlead that couples your TV set to its rabbit-ears, or the 52-ohm coaxial cable used to connect a ham or CB transceiver to its antenna.

You'll recall that we've repeatedly emphasized that the concept of impedance was, in the most general case, just a measure of the ratio of voltage across a circuit to current in that circuit, and that any expression of such a ratio would be measured in ohms.

We can take this a step farther, and call it a ratio of **force** to **response**. It then becomes the electrical equivalent of mechanical elasticity. Going back to Maxwell's theory as described in Chapter 1, we can call it a measure of the compression-to-expansion ratio.

Since free space will support electromagnetic radiation (if it couldn't, we would receive neither light nor heat from the sun!), it must be capable of both compression and expansion if the theory is correct, and this would mean that we could define an impedance for free space. Believe it or not, that has been done—the value is 377 ohms!

The exact figure is not important. What is important is the extension of the idea of impedance from a **hard circuit** in

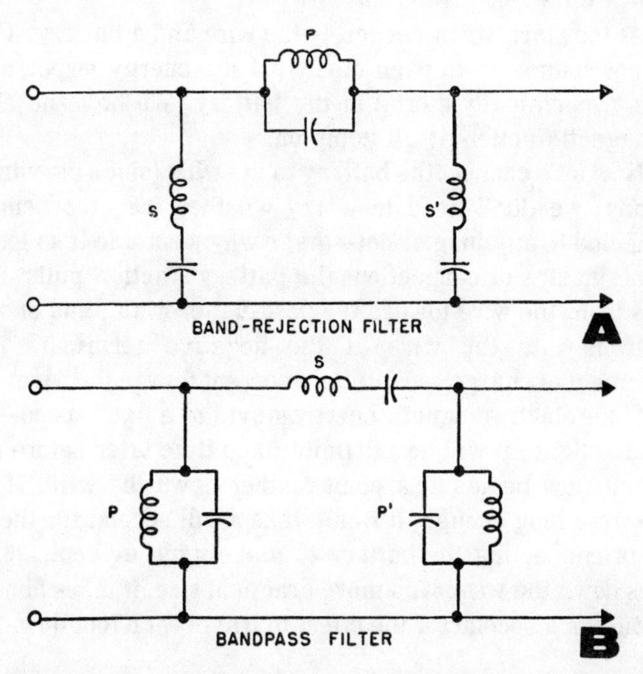

Fig. 20. These filters are complementary to one another. The bandstop (or band-rejection) filter short-circuits signals at a specific frequency; the bandpass filter allows only signals at a specific frequency to pass.

which all elements are connected by conductors, to the area of **radiated energy.**

The 377-ohm impedance of free space is obtained directly from two quantities called **permeability** and **dielectric constant.** Permeability is a measure of the expansion capability, and dielectric constant is a measure of the electrical compressibility.

The same two properties can also be expressed in terms of inductance and capacitance, since inductance is closely related to permeability, and capacitance to dielectric constant.

Now let's take a look at a transmission line. Just to make it specific, let's look at 300-ohm TV twinlead.

While we're at it, let's imagine that our twinlead is long. It doesn't really matter how long, because we're only going to be concerned with what happens at a specific point on the line, but it's a little easier to imagine what goes on as being isolated from everything else if we think of it as happening part way down a mile-long stretch of wire.

At the start, we have merely the wire and a battery. They are not connected to each other. All the energy is potential energy, chemically stored in the battery. We have no electromagnetic action at all going on.

Now let's connect the battery to the wire (since the wire is so long, we don't need to worry whether the other end is connected to anything or not—that's why we made it so long). At the instant of connection, the battery's action pulls electrons from the wire toward the positive terminal, and shoves electrons into the wire at the negative terminal. This movement of charge constitutes a current flow in the wire.

Since electromagnetic energy moves at a fixed speed—the speed of light—it will be a definite fixed time later before any current flow begins at a point farther down the wire. If the wire were long enough, it would take a full second for the effect of connecting the battery to make itself evident 186,000 miles down the wire. At a more practical size, it takes about a billionth of a second for the effect to travel each foot down the wire.

As the current "front" or "step" moves down the line, current must continue to flow behind it also at the same rate.

We can think of this current as a capacitor-charging current, flowing to charge the capacitance between the two wires of the line.

As the current flows, it is accompanied by a magnetic field which induces opposition to the current flow, so the line has inductance as well.

The major difference between the capacitance of the line and that of the capacitors we looked at back in Chapter 3 is that in the line, the capacitance is distributed down the entire length of the line; in the capacitor, it's lumped into one element.

The same difference distinguishes the line's inductance from that of the inductor. Thus we say that the reactances of the transmission line form a **distributed constant**, while those in a conventional resonant circuit are composed of **lumped constants.**

Now we have seen that inductance is a measure of response and that capacitance is a measure of force, which means that impedance can be defined in terms of a ratio between L and C. The actual definition is that the so-called characteristic or **surge** impedance of any medium through which electromagnetic energy travels is equal to the square root of the ratio L:C, where L and C are inductance and capacitance per unit length. This is how the 300-ohm impedance figure for twinlead is determined in the first place. L depends on wire composition and diameter to a large degree, and C depends on conductor diameter, dielectric material, and spacing between conductors.

The same principles apply to the impedance value for coaxial cables, but the formula is different because the field distribution differs.

Note that the line impedance does not depend upon the circuit being complete at the far end, because it's determined entirely by what happens as the current step moves down the line.

Any place along the line that the characteristics change, though, the current step requirement will be different. This will, in turn, cause some of the energy to be reflected back toward the battery. The simplest example of this is what happens at the end of the line if it ends in an open circuit. The current can go nowhere else, so must stop flowing. However, because of self-inductance, it cannot stop instantly. What happens instead is that it reflects right back toward the battery, but with reversed polarity, to cancel incoming current flow.

The result is a multiple reflection of energy back and forth along the line, until the situation stabilizes. At that time the conditions at the open end of the line duplicate those at the battery terminals, no current is flowing, and everything is back in balance.

Something similar would have happened had the line merely connected to another line of different impedance. The only way that the current step can move smoothly forward, without any energy being reflected back toward the source, is to stay on the same line—or at the same impedance level, because the energy can only be affected by impedance-level changes.

That's what makes impedance matching so important in rf transmission lines. So long as the line is connected to a load of the same impedance level, the energy cannot tell the difference from an infinitely long line, and no reflections occur. When impedances are not matched, each mismatch creates energy reflections, and efficiency of the line is reduced.

Active Circuit Elements

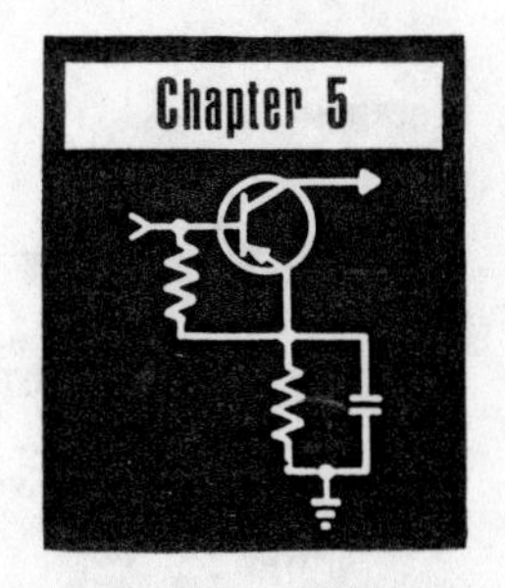

In Chapter 3 we became acquainted with the most common passive circuit elements, and in Chapter 4 we learned how these passive elements can be combined into passive circuits. Most actual circuits we will encounter in practice, though, involve more than just passive elements; they also include elements we term **active**. In this chapter, we'll meet the most common active elements.

The difference between an active and a passive element can be stated simply. A passive element's characteristics—its resistance or reactance, or its conductance or susceptance—are determined by its construction or by its adjustment, in the case of variable or adjustable elements. In no case, though, is any key characteristic changed by circuit action. An active element provides some characteristic which is changed by circuit action. That's the reason the element is included in the circuit, and this capability is what makes electronics the powerful technology it is today. Once a circuit has the capability of changing its own characteristics, it becomes possible to build circuits which can give frighteningly close imitations of intelligent activity.

The circuits we will examine in this volume come nowhere close to this level of complexity, but even the most complicated computer circuits are made up of basic building blocks, each of them simpler than most of the circuits we're seeing here.

THE RELAY

The oldest active device employed in electrical circuits is actually electromechanical in nature, rather than electronic, but is presented here to introduce the concept of active elements as well as to introduce the device itself. This element is the relay, which dates from the middle of the 19th century and the golden age of telegraphy.

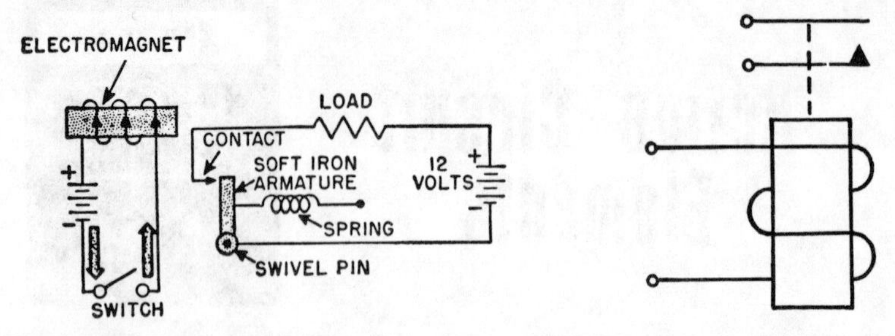

Fig. 5-1. Schematic symbol for this relay, at right, is similar to arrangement of parts in relay itself, left. Sometimes the relay's core is shown by dotted lines alongside the coil, and ganging of contacts may be indicated by connecting the wiper arms by dotted lines. If only one relay appears in a circuit, however, such frills are often omitted. The relay shown is a single-pole single-throw type comparable to an SPST switch of the **momentary-contact** type.

A relay consists of an electromagnet coupled to switch contacts, so that the switch is operated automatically when the electromagnet is energized. Thus, application of power to the relay coil causes the contacts to operate, changing the impedance across the contacts from high to low (or vice versa), depending upon the contact arrangement.

Some relay designs emphasize operation from low-current, relatively high-voltage sources, while others appear to be current-operated. Actually, the operating force for all relays is power. If the coil is wound with many turns so that it has high impedance, low current will operate the device; if the coil contains fewer turns with low impedance, low voltage will suffice but the power must be made up with higher current.

The coupling between the coil and the contacts is mechanical. This limits the speed of response of any relay. The fastest-acting relays available require several thousandths of a second (milliseconds) for action, and even then the contacts "bounce" for several milliseconds after the first action. Designs which minimize contact bounce are slower in their initial action. For this reason, relays are used primarily to control power and operate logic signal lines; they are never used to amplify audio frequencies.

A typical relay is shown schematically in Fig. 5-1. When the coil of the electromagnet is inactive, the soft iron armature is pulled from the contact by a mechanical spring. Actuating the coil by closing the switch reverses things. The electromagnet, once energized, overcomes the pull of the spring and attracts the armature, which then touches the contact, thus closing the 12V circuit and applying power to the load. This type of relay has an action very much like an SPST switch. Most relays are active only so long as power is applied. Mechanical latching relays are available, though, which serve as the equivalent of the toggle or knife switch that is actuated once and remains in its new position. These relays, however, often require a second coil for unlatching; they are not common, for that reason.

Figure 5-2 shows a circuit which provides SPDT action with DPDT contacts. When switch S is closed (as shown), battery power is supplied to the relay contacts, but the coil is not energized. When the momentary-contact PULSE switch is closed briefly, the two wiper arms are pulled in toward the armature, thus closing two circuits: the C-NO of the top set and the equivalent two on the bottom set. Battery voltage is now applied to the coil through the lowest contacts even though the momentary-contact switch is no longer closed. Opening the S switch is the only way to remove power from the circuit; doing this will release the closed contacts to reset the relay to the condition shown.

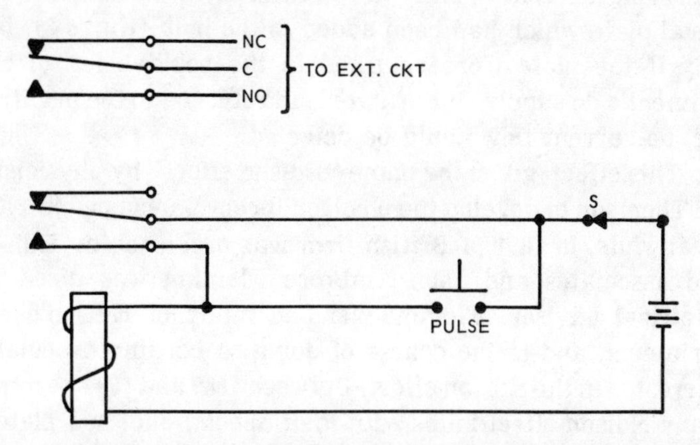

Fig. 5-2. This circuit provides latching action for any relay which has a spare contact available. The PULSE pushbutton makes the relay close momentarily. But even when the voltage pulse is gone the relay stays closed; the relay's own lower contacts provide a constant coil voltage. Breaking the circuit at S resets the relay. This circuit may be said to "remember" the fact that an input pulse was applied; removal of power (opening S) is then equivalent to an order to forget this memory. Computer arithmetic registers are only slightly more complicated than this.

Many other relay circuits are possible. For example, the first electrical computers built by IBM and the Bell Telephone System were composed of hundreds of relays. The only major problems were power consumption and relatively slow speed.

THE VACUUM TUBE

If any single circuit element can fairly be said to be responsible for the growth and acceptance of electronics as we know it, the vacuum tube is that element. For nearly half a century this device formed the backbone of electronics; the electron theory which gave our art its name was developed in an effort to determine what goes on in a vacuum tube, and even today both radio broadcasting and television are completely dependent upon the tube for their existence.

The vacuum tube was developed as a direct outgrowth of Edison's incandescent lamp. When that first electric lamp's 40-hour life came to an end on October 21, 1879, the inside of its glass envelope had been blackened. In his efforts to extend bulb life and improve the device Edison performed many experiments. He determined that more than light and heat was being emitted from the bulb's filament, and in his notebooks dated February 1880 is the first mention of "carrying current." This was a current which flowed to a metal plate which had been added to the bulb (Fig. 5-3), but only if the plate were returned to the positive leg of the filament's dc supply. If the plate lead returned to the negative leg, no current flow could be detected.

This effect, given the name "Edison effect" by physicists, led Thomson to develop the electron theory announced in 1900. Meanwhile, in 1881, a British firm was organized by Edison and associates and John Ambrose Fleming was hired as electrical adviser. Fleming verified many of Edison's experiments, and in the course of doing so became especially interested in the Edison effect. Between 1889 and 1896 he made many Edison-effect bulbs with their special enclosed plates, and determined that the "carrying current" followed definite rules. Strength of the current flow depended upon the distance between filament and plate. The current flow could not turn corners, but moved only in a straight line between filament and plate. Placing a thin sheet of mica (an insulator) between filament and plate stopped the flow, proving that the plate could be screened. And, most important of all at the time, Fleming learned that current could flow only one way through the device. When the plate was negative, current never flowed. Edison had noted this in his first observations, but Fleming put it to the test by using it as a "valve" to change ac to dc.

By 1896, Fleming felt that he had learned all he needed to know about the Edison effect, and the experimental bulbs were stored away in a cupboard. Three years later, Marconi entered the scene when Fleming became a technical advisor to the radio pioneer. Fleming was given the job of improving the receivers used at the time, and he recalled his success at

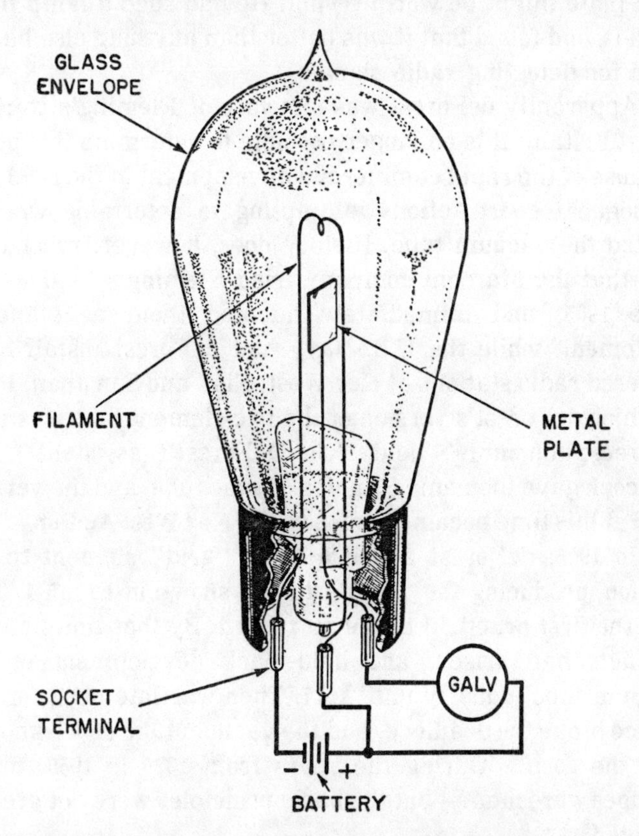

Fig. 5-3. This is the granddaddy of all vacuum-tube circuits. This circuit, or one very like it, was used by Thomas A. Edison in the first 6 months of the electric light bulb's existence to determine presence of the **Edison effect**, which led to the vacuum valve used by Fleming and the Key West Audion employed by deForest, and thus evolved into the modern vacuum tube.

"rectifying" low-frequency ac into dc with the Edison valve. Hastily digging one of the old bulbs from storage, he tried it. "The experiment was at once a great success," he wrote later. Thus the first application of the vacuum tube to practical electronics was made in October 1904.

Meanwhile in the United States, Lee deForest had become convinced that in some fashion, incandescent gases could be used to detect radio signals. In 1903, he built a "flame detector" using two platinum electrodes in a gas flame, and picked up signals from ships in New York harbor. Edison had discovered that the blackening of the electric light bulb could be reduced by filling the bulb with gas, instead of using vacuum, and in 1905 deForest thought that an electric light with plate might be worth trying. He had such a lamp built, tried it, and found that it was better than anything else he had used for detecting radio signals.

Apparently deForest was unaware of Fleming's work in Great Britain. It is no longer possible to determine this point, because of the rapid commercial development in the field and subsequent court actions attempting to determine who invented the vacuum tube. History does, however, record the fact that the Marconi company tried Fleming's "valves" in June 1905, and immediately adopted them as standard equipment, while the U.S. Navy had deForest install high-powered radio stations at Key West, Fla., and San Juan, P.R., in which deForest's version of the two-element tube was used for reception in 1904 and 1905. DeForest's assistant C. D. Babcock gave the name "Audion" to the tube, and the version used at this time became known as the **Key West Audion**.

In 1906, deForest added a third "grid" element to the Audion, producing the "grid Audion" shown in Fig. 5-4. This was the first practical triode on record. By that time, patent conflicts had arisen, and held back development of the vacuum tube industry until 1917. Then wartime need for the device broke the deadlock, and the vacuum tube as we know it was the result. During the years from 1920 to 1950, many changes were made—but the basic principles were not greatly affected.

The Vacuum Valve—The Diode

The original Edison-effect bulb, refined into the Fleming valve and deForest's Key West Audion, consisted only of a filament and a metal plate, sealed inside a glass envelope which made it possible to reduce the atmosphere surrounding them to vacuum. Since it had only two elements, it came to be known as a diode (a **blend** word created from **di**, meaning **two**, and **electrode**).

Today's vacuum diode differs from those first models in principle of operation; the 75 years of development have produced only such distinctions as increased ruggedness, longer life, higher vacuum, improved materials of construction, etc.

Like all vacuum tubes, the diode owes its operation to a cloud of electrons emitted from the heated filament or **cathode** of the tube, which then travel through the vacuum between cathode and plate or **anode** to permit current flow through the tube. The starting point in understanding its principles, then, is to determine how it happens that the cloud of electrons is emitted.

In Chapter 1 we saw that, according to electron theory as elaborated by presently accepted rules of physics, the difference between conducting and insulating properties of various materials depends upon the occupancy and spacing of the energy bands at the atomic level of structure in the

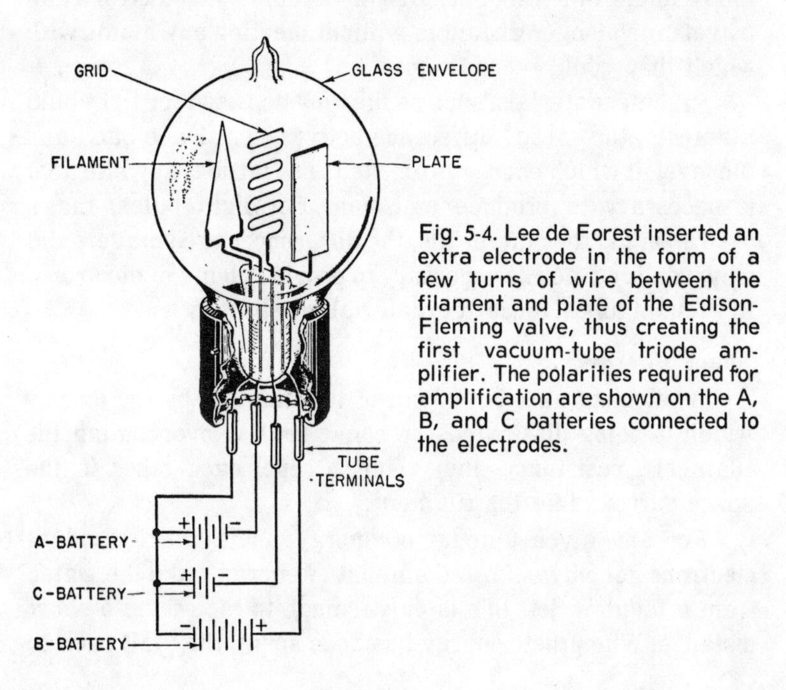

Fig. 5-4. Lee de Forest inserted an extra electrode in the form of a few turns of wire between the filament and plate of the Edison-Fleming valve, thus creating the first vacuum-tube triode amplifier. The polarities required for amplification are shown on the A, B, and C batteries connected to the electrodes.

material, and that when energy is added to any material in the form of heat, the spacing of these energy bands is affected.

An important point is that it is temperature, rather than the mere quantity of heat, that determines what happens to the energy levels and to the electrons. Physicists now believe that this temperature or energy measure is directly related to frequency. Thus, heat is higher in the frequency spectrum than is radio energy; visible light is higher in frequency than heat. X-rays and cosmic rays, in turn, are of even higher frequency. As the frequency increases, so does the energy, all other things remaining constant.

That's why when Edison put enough energy into the filament of his light bulb to cause it to get hot enough to give off useful light, other things were emitted as well. At visible temperatures, some of the energy kicks applied to individual electrons can be hard enough to knock them right past all the vacant energy levels, out away from the atomic structure into free space. In air, they can rapidly recombine with other atoms, as their places back in the filament structure are taken by valence electrons of oxygen and the filament burns up. This is why Edison got nowhere with the electric light until he put the filament in a vacuum! But in vacuum, the electrons can travel much longer distances without meeting any atoms with which they could recombine.

In some materials, such as the tungsten used for light-bulb filaments, the energy difference between a valence band and the level at which energy is emitted is so large that white heat is necessary to produce emission. For light bulbs, that's desirable. In other materials, the differences are smaller, and only a dull red glow is necessary to provide plenty of electrons. In vacuum tubes, the latter materials are usually used.

Space Charge

As electrons are boiled out of the filament by the energy which is being dissipated (or converted) in overcoming the filament's resistance, they tend to repel each other in the space surrounding the filament.

For any given amount of energy input, the individual electrons get only a limited amount of energy. Like the water from a fountain jet, this is only enough to take them a short distance. When their energy has been spent, they fall back to the emitting surface. (See circles with arrows in Fig. 5-5A.) A few, of course, will get through to break free of the attraction of the filament and will shoot off to infinity as X-rays, but at normal energy levels these are so few that we can safely ignore them (less radiation results from this than we receive naturally in the form of cosmic rays).

The resulting situation is that we have, in the vacuum surrounding a hot filament, an energetic cloud of boiled-off electrons which have been emitted from the filament and have not yet fallen back to its surface, and which repel each other. They have no place to go, and so merely stay where they are. Since they are all electrons, they represent a definite and measurable amount of electric charge. We call this the "space charge" surrounding a heated cathode.

When Edison, Fleming, and deForest put their metal plates into the bulbs where the space charge existed and completed a circuit between plate and filament, the electrons of the space charge were able to flow across the vacuum to the

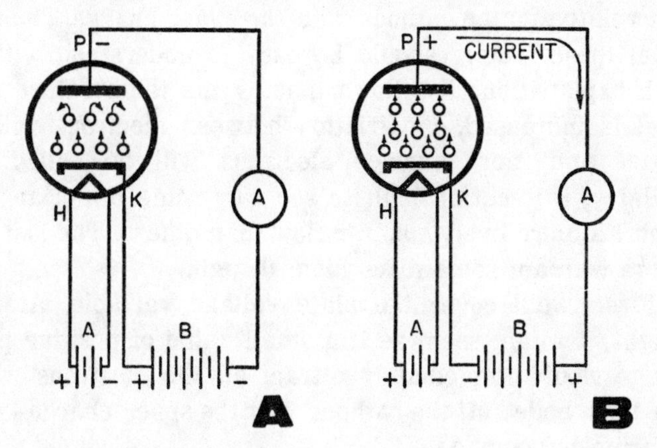

Fig. 5-5. Diodes manufactured in recent years have a filament or heater (H) that serves only as a "stove" to evenly heat the surface of a cathode (K). In A, the plate (P) is connected to the negative terminal of a battery through an ammeter (circled A). The electrons themselves are negatively polarized, so the plate repels the electrons, and the electrons repel each other. The profusion of electrons in the space charge just "loiter" because they have nowhere to go. However, when a sizable positive potential is applied to the plate, as in B, an electrostatic attraction is set up and the electrons furiously migrate from the cathode to the plate. The current registered on the meter will be the same as the current flowing in the cathode circuit. This is because the cathode circuit is actually a part of the anode circuit. If the cathode connection is broken, there is no longer a complete path for electrons to travel through the circuit.

plate, and then through the external circuit back to the filament.

When the plate circuit is supplied with a negative voltage as in Fig. 5-5A, the boiled-off electrons just "loiter"—they do not accumulate on the plate because the negative potential actually repels the space charge. But when the plate is made positive with respect to the filament or cathode, the electrons in the space charge are attracted rigorously to the plate surface, as shown in Fig. 5-5B. This **attraction** of the plate sets up a current flow in the plate circuit, as shown. By now it should be obvious that the cathode or filament of an electron tube is an **emitter**. The anode or plate is a **collector**. The cathode **emits** electrons and the plate **collects** them. These terms are important to remember when we discuss semiconductors.

The degree of attraction of the plate for the emitted electrons depends on several factors: the **temperature of the cathode**, the **emission per unit time** (more simply, the material used for the cathode element), the **voltage of the plate** relative to the cathode, and **the space charge.** The first two of these factors should be easy to understand without much explanation: it follows that as the temperature of a metal is increased, the friction between electrons will be greater and more surface electrons will be boiled off. Similarly, it is not difficult to see why some materials will support a more lively electron flow than others. The last two factors warrant some discussion, though.

First, we'll cover the plate voltage variable. At first thought, it might seem as if a small value of positive plate voltage would immediately attract all the electrons which have been boiled off the cathode. But the space charge keeps this from happening.

Imagine for a moment a tube arranged as illustrated in Fig. 5-6A. The cathode is heated to an electron-emitting temperature and the plate is joined directly to the cathode. Since no voltage is applied to the plate, it can exert no attracting force on the electrons coming off the cathode. So what happens to these "loitering" electrons?

Electrons don't really boil off in ranks, as shown in B, but it helps to visualize the process this way, so we will. We'll start with the emission of the first row of electrons. They have nothing in front of them to impede their progress, but then again they aren't being "shot" off the cathode in the first place—it's more of a "liberation." And there are other electrons being boiled off directly behind them that have a like charge and thus repel those electrons on our first row. The electrons in the second row (behind the first) are being repelled by the electrons in the first row, too, because the repulsion is fully mutual. The net result is a total retardation of movement, as layer upon layer of electrons feel these forces

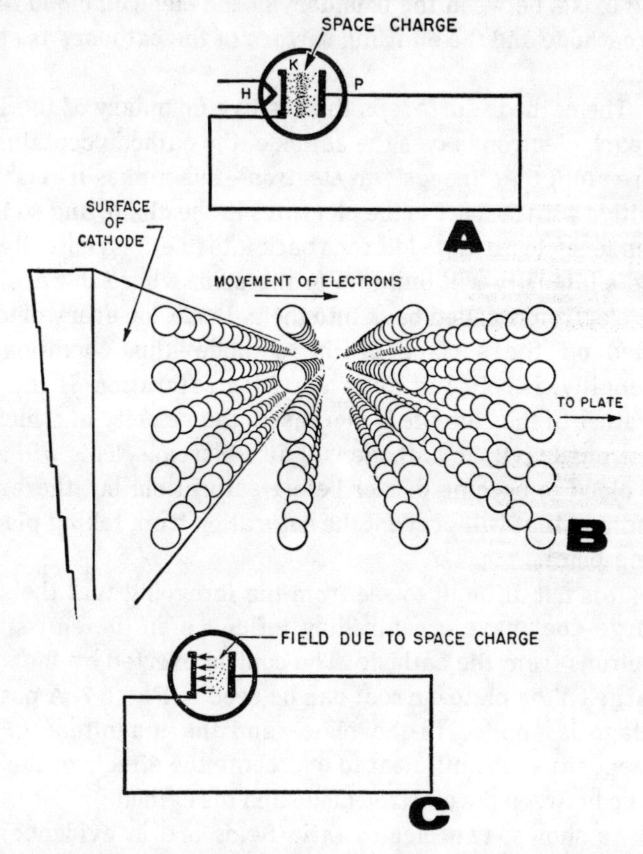

Fig. 5-6. When the cathode is connected directly to the plate (A), an applied heater voltage causes electrons to be boiled off the surface of the cathode. But these electrons are being repelled by other electrons following them from the cathode surface while the cathode itself is building up an affinity for electrons (because of the deficiency caused by the boiloff). The result is an army of electrons that form into a cloud (B). The repelling force of other boiled off electrons keep the cloud from being attracted back to the cathode as an entity, but an electrostatic force does build up between the electron cloud (space charge) and cathode (C).

acting on every surface, and this sluggish action serves to allow a buildup of large quantities of electrons in the general vicinity of the cathode. These electrons actually are the cloud we refer to as the "space charge."

The cloud is quite dense near the cathode, but less dense away from it, owing to the fact that the extreme-distance electrons have lost much of their energy and are still being repelled by the "mass." Because it is indeed a cumulative charge, there is an electrostatic field associated with it. This field exists between the boundary of the electron cloud facing the cathode and the emitting surface of the cathode, as shown in C.

The cathode surface is the positive boundary of the field. As each electron leaves the cathode, the cathode contains one more atom that "needs" an electron—this makes it relatively positive with respect to the electrons in the cloud, and so it has a tendency to attract electrons back into itself. Eventually, the field's intensity will build up to the point where there will be one electron repelled back into the cathode for every electron boiled off the surface of the cathode—this condition, incidentally, is referred to as **emission saturation**. If the temperature of the cathode is increased, the velocity at which the electrons leave the cathode will also increase. This will allow the cloud to become denser before saturation, but there is no condition that will prevent the saturation from taking place at some point.

It is not difficult to see from the foregoing that the space charge does have a controlling influence on the emission of electrons from the cathode. The control exerted by the space charge on the plate current can be seen in Fig. 5-7. A positive voltage is applied to the plate—and the magnitude of that voltage must be sufficient to overcome the effects of the field set up between the electron cloud and the cathode.

As shown, two electrostatic fields are in evidence: one between the cathode and the space charge and the other between the far end of the space charge and the plate. Because of the presence of the space charge, it is proper to view the attracting force of the positively charged plate as acting on the electrons in the space charge rather than directly on the individual electrons being emitted. This will help in visualizing the process that's really happening: It isn't a matter of each electron leaving the cathode and being shot directly at the plate; the cloud is always there. The electrons boil off the cathode at one velocity and gather in the cloud. At the same time, electrons at the other side of the cloud are suddenly and with force attracted to the plate. The process is distinctly a three-step affair: (1) cathode to cloud, (2) migrate across cloud, and (3) cloud to plate.

Characteristic Curves

If we vary the plate voltage across a wide range of values, taking plate-current measurements all the while, then plot a graph of the results, we will come up with something that looks like Fig. 5-8. Such a graph is called the **characteristic curve** of the tube. Normally, the graph will have several such curves, each curve based on a different fixed value of heater voltage. The exact performance of any tube can be predicted from a set of such curves.

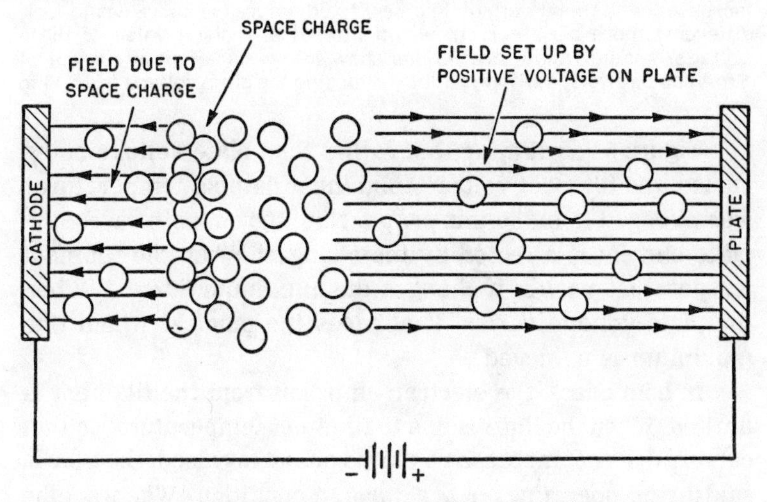

Fig. 5-7. The fact that two electrostatic fields exist in the space between the cathode and plate of a vacuum tube diode means that an electron goes through three steps (and velocities) as it makes the journey from cathode to plate. The first is a battle between the negative repulsion of the electrons in the space charge and the stronger pushing force of the electrons being emitted—all this taking place in the vicinity of a cathode that keeps pushing out electrons while attracting them because of its positive-charge buildup. The second is the mutual repulsion of the electrons in the space charge, as each electron makes its way through the cloud. The third is the sudden buildup of velocity as electrons push from behind and the positively charged plate pulls from ahead.

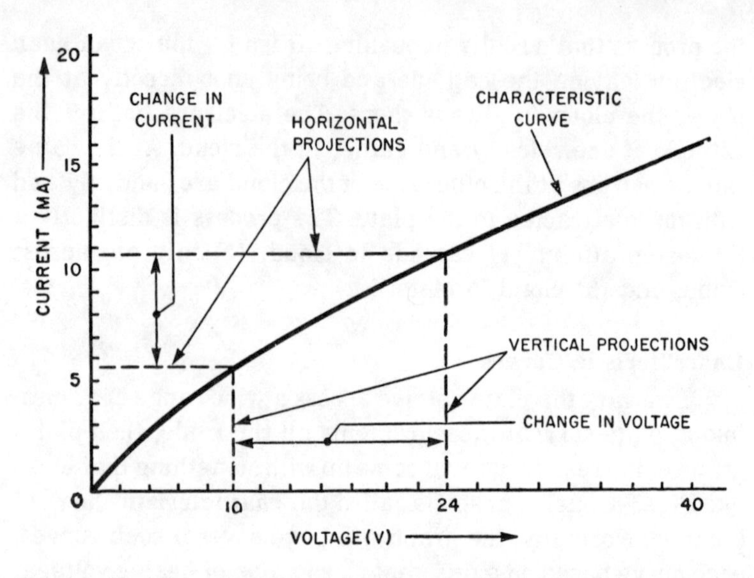

Fig. 5-8. If plate voltage is continuously varied while plate current is measured, and resulting measurements are plotted on a graph, a typical diode might produce the curve shown here. This curve results from doing just this with the filament voltage set to a constant value. If we were to increase the filament voltage to a new fixed value, the curve would look different; more plate current would flow at any given value of plate voltage. The horizontal projections show that a variation in current of 5 mA can be effected with this tube by varying the plate voltage from 10 to 24V.

Figure 5-9 is similar; but rather than plate voltage being varied, the filament voltage (and thus filament temperature) was changed to make this graph. Note that in both cases, the plate current is affected by the changes. When the filament temperature varies, it changes the maximum current. When the plate voltage varies, it changes the point at which this maximum is achieved.

In both cases, the electron emission from the filament is limited. When the limit is due to filament temperature, so that current does not increase as plate voltage is raised, the tube is said to be operating in a **saturated** condition. When a tube operates in saturation, one electron is absorbed from the space charge by the cathode for every electron emitted by the cathode—it's a state of equilibrium.

Neglecting the space charge, the circuit between filament (or cathode) and plate of a vacuum tube can be thought of as a resistor, up to a point. As voltage across it increases, so does the current. By voltage-divider action, the voltage across any part of the resistor is proportional to the fraction of the total resistance represented by that part. This "ideal" situation never occurs in practice, but the concept it represents is vital to an understanding of how vacuum tubes—and any other active elements—work.

In normal practice, plate voltage is applied but the tube is not saturated. This means that some, but not all, of the emitted electrons go to the plate, so that we have plate current. That part of the circuit corresponds to our resistor of the **perfect-saturation** case. The rest of the electrons form a space charge, which acts like a negative battery in the middle of the resistor.

The result is that the voltage gradient between cathode and plate is not actually smooth and straight. Instead, it has a kink in it which corresponds to the effect of the space charge.

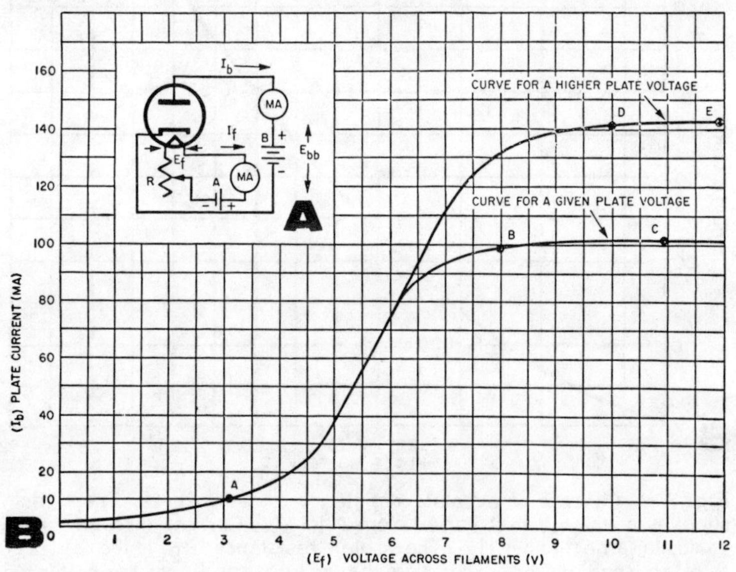

Fig. 5-9. Any tube characteristics which may be varied can be changed to plot characteristic curves. Here, for instance, are curves showing how plate current varies as filament voltage is changed, for two fixed values of plate voltage. While information is essentially the same as that shown in Fig. 5-8, the emphasis is different. Note how this curve makes it obvious that plate voltage has strong effect on current, while curves of Fig. 5-8 bring out dependence on filament voltage and limiting effect of low voltages on filament. The symbols in the sketch are as follows: I_b = current intensity from the B battery, or simply plate current; E_{bb} = electromotive force (voltage) from B battery; I_f =filament current; E_f= filament voltage; A ="A" battery (traditionally so called to distinguish low-voltage batteries from high-voltage batteries); R = resistor (used to vary filament voltage); MA = meters for monitoring current, in milliamperes.

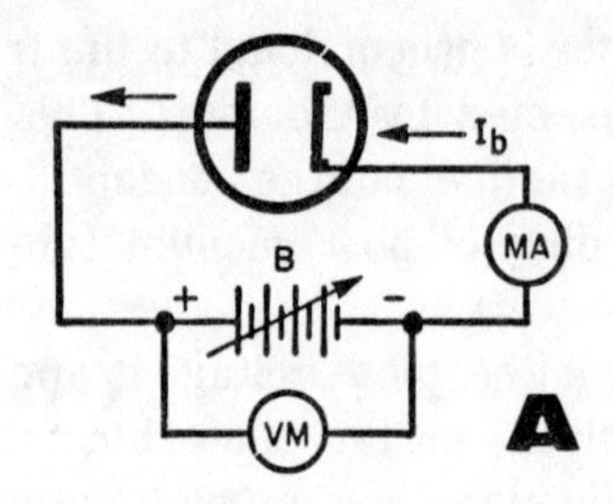

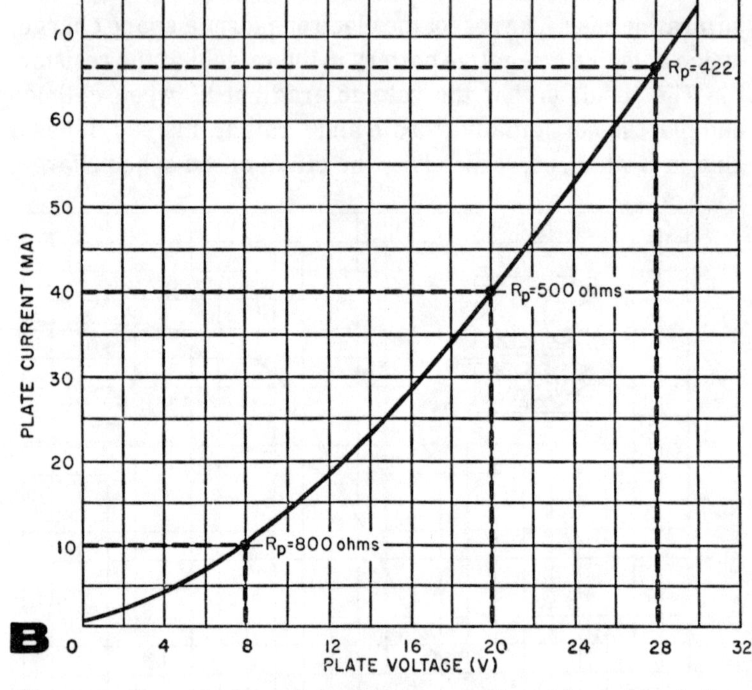

Fig. 5-10. Concept of ac plate resistance is shown here. The battery supplies a voltage that varies from 0 to 32V Current through circuit varies in same manner (I_b). The ac plate resistance (R_p) is the ratio E: I, as established by Ohm's law. Note that ac plate resistance has nothing to do with actual dc voltage and resulting average tube current, but depends only upon the changes. Observe the values of R_p at the three plotted operating points.

If plate voltage is lower than the effective voltage of the space charge "battery" no current can flow. If plate voltage is only slightly greater than the effect of space charge, only a small current flows. If plate voltage is much greater, the current is also very large and the effect of the space charge virtually disappears.

The result of all this is that the vacuum diode acts like a resistor with one-way current flow (electrons cannot move from plate back to cathode), but the curvature introduced into any tube's characteristic by the effect of the space charge causes the resistance to be **nonlinear**—that is, where a linear 5000-ohm resistor is 5000 ohms at 5V or at 500V, a nonlinear resistance such as that of a diode may be 500,000 ohms at 5V and only 500 ohms if the voltage rises to 500V. Fortunately, for relatively small swings of voltage or current, the tube's characteristics can be made essentially linear and the similarity to a resistor can be used to advantage.

We even speak of the "ac plate resistance" of a diode, and when we do so we're talking about the effective value of resistance it represents to small changes in signal level. Figure 5-11 shows how this works. If we apply 20V dc to a diode

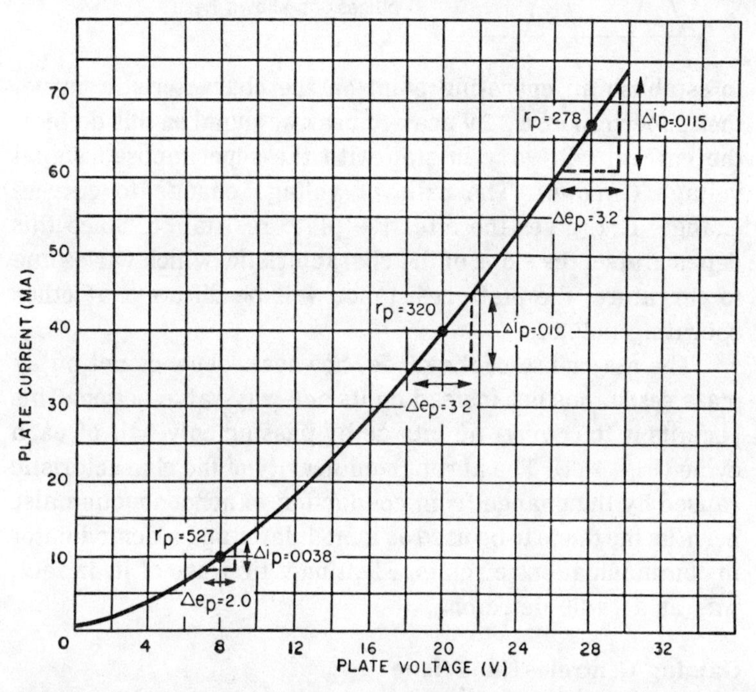

Fig. 5-11. The characteristic curve can be used to determine the ac plate resistance of a diode. Look at the 20V operating point. Varying the voltage here by applying a 3.2V p-p signal (horizontal arrows) gives a plate current variation of 10 mA. At the operating point, the plate resistance is 320 ohms. At a plate voltage of 28V, the same 3.2V variation results in a plate current swing of 11.5 mA. The operating point plate resistance in this latter case is 278 ohms. In each case the operating point plate resistance is determined from Ohm's law: R = E/I.

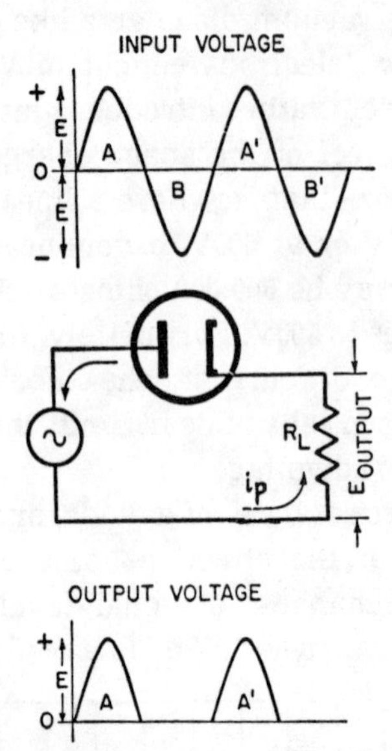

Fig. 5-12. It isn't hard to understand how the diode rectifies if you remember the characteristic curve. The input voltage increases from zero and the diode conducts accordingly. As the input signal (A) peaks, then declines, the tube continues to conduct, as predicted by the characteristic curve. However, when the input signal returns to zero and starts the reverse-polarity excursion (B), the situation is identical to that of the diode with a negative plate voltage: no plate current can flow. The tube's output, then, consists of a series of positive pulses, as shown here.

to establish an operating point on the characteristic curve, then superimpose a 3.2V peak-to-peak ac signal on this dc **bias**, the current will vary in step with the superimposed signal voltage (arrows). The ratio of voltage change to current change, E:I, gives the effective plate resistance. Since this depends upon the slope of the characteristic, which varies due to curvature, the plate resistance will be different at other operating points.

The major use of the diode, however, depends not on its plate resistance but instead on its one-way-valve action. This permits it to convert ac into dc by passing only half of each cycle (Fig. 5-12). The abrupt nonlinearity of the characteristic caused by the change from conduction to nonconduction also permits the diode to be used as a modulator and a demodulator in communications circuits. Fleming's first use of it, in fact, was as a radio detector.

Gaining Control—The Triode

When Lee deForest added a third electrode to his **Audion** in 1906, thus becoming the inadvertent inventor of the triode tube (three electrodes), he was simply trying to improve the diode's action as a radio detector. His first experiment was to wrap a diode in tinfoil and connect the antenna to the foil wrap.

This arrangement worked better than the conventional diode, so his next step was to make a diode with two separate plates. He connected the antenna to one and the headphone to the other. This also worked well, but he felt that closer coupling would be still better.

Remembering that any solid barrier between cathode and plate would block current flow, he bent wire in the shape of a grid and inserted it between filament and plate. This tube performed best of all, and the "grid Audion" was born.

Not until some time later did anyone discover just what the grid was really doing in the tube. Only after the tube's capability of amplifying signals become apparent did its theory of operation become known.

If we insert a grid between cathode and plate of a vacuum tube, but leave it disconnected from both cathode and plate, the grid will have no effect at all on the electrons traveling from cathode to plate. Suppose, though, that we connect the grid externally to the cathode so that the grid's potential is that of the cathode, as shown in Fig. 5-13. In this case, some of the electrons flowing through the screen will strike the grid and start a small current flowing in the grid circuit. Grid current is not desirable, for it would not allow us to use the grid for other purposes, which we'll explain later. Most of the electrons, though, would pass right through the grid structure to be attracted to the plate electrode.

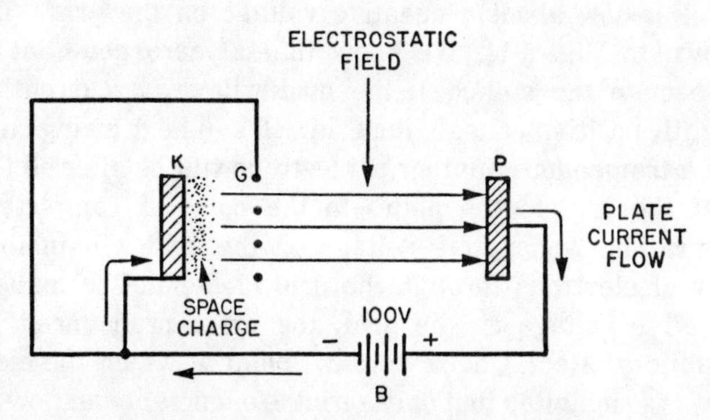

Fig. 5-13. The grid (G), represented by the four vertical dots in this sketch, is connected directly to the cathode (K). Now the grid serves as a crude fence; some of the electrons are trapped in the grid structure as they make their way across to the plate. When this happens, grid current flows; the current is too small to be useful, and is actually an undesirable factor in most circuits.

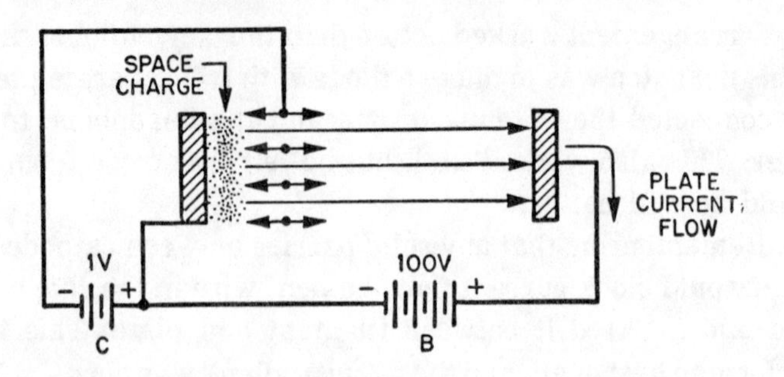

Fig. 5-14. When the grid is **biased** slightly negative, as shown here, it repels electrons. This keeps errant electrons from landing on the grid and thus causing current flow where it's not wanted. The more negative the grid bias, the greater the repelling force. If the negative grid voltage is high enough, all the electrons leaving the cathode are repelled right back without even reaching the grid. When this occurs, the tube is said to be "cut off."

Applying a positive potential on the grid has a very interesting effect on the space charge. Since the grid is situated in fairly close proximity to the cathode, a slight positive voltage will serve to help pull the electrons from the area adjacent to the cathode. As the electrons reach the grid, they are well out of the influence of the cathode's attraction because of the strong positive field at the grid. If the plate voltage is high enough, it can easily overpower the influence of the grid, and pull electrons right through the grid of the plate. So much does the grid aid the flow of electrons, in fact, that the tube goes into saturation immediately.

But how about a negative voltage on the grid? This is shown in Fig. 5-14. We know that at zero potential (with respect to the cathode), the grid will draw current. At a slightly positive voltage, the grid will still be drawing current, but a tremendous number of electrons will be allowed to flow past the grid to the plate—to the point of complete tube saturation. A negative voltage on the grid will inhibit the flow of electrons through the grid element. The higher the negative voltage on the grid, the more pronounced is the inhibitory effect. There will be a point at which the electron flow is so inhibited that only a trickle of current can flow to the plate.

This is the basis for signal amplification.

We have seen that the plate resistance of a tube depends upon the characteristic curve at the desired operating point.

We have also seen how changing the grid voltage effectively changes the shape of the voltage gradient within the tube; and since the characteristic curve is just one effect of the voltage gradient, any change in grid voltage must change the characteristic curve.

That means that changing the grid voltage of a triode must change its plate-to-cathode resistance. Another, and much simpler way of saying this is to declare that **plate current depends upon grid voltage.** However, to get a good idea of the similarities among all types of active devices, it's best to think in terms of a resistance change controlled by grid voltage.

It would probably be a good idea to go over these points once more for yourself, because they are extremely important in your ultimate understanding of electronics. Examine Fig. 5-15, which has a characteristic curve (A) and a triode circuit (B). Note in the circuit that the grid of the tube can be supplied with a varying voltage of either positive or negative polarity.

In the graph on which is plotted the characteristic curve of this circuit, grid voltage is plotted across the bottom, and plate current is registered on the vertical scale. In this case, when the grid is at +4V, the tube saturates; further increases in grid voltage have little effect on this current. As the grid voltage is lowered, the characteristic curve begins to look less like a curve and more like a straight line. This is particularly true as the grid voltage passes through the zero point. This region of the curve below the zero point is referred to as the **linear** portion of the characteristic. If the grid is made even more negative, plate current drops substantially, eventually cutting off completely at about —6V.

You can see what would happen in the plate circuit of this triode if the value of grid voltage were varied rapidly between, say, —2.7V and —0.7V; the plate current would vary about 14 and 30 mA.

Actually, the operation of the triode is quite similar to the action of the relay. There, resistance between relay contacts was controlled by power through the relay coil. True, the choice was between very high resistance (open circuit) and very low (closed contacts), but it was definitely a change in

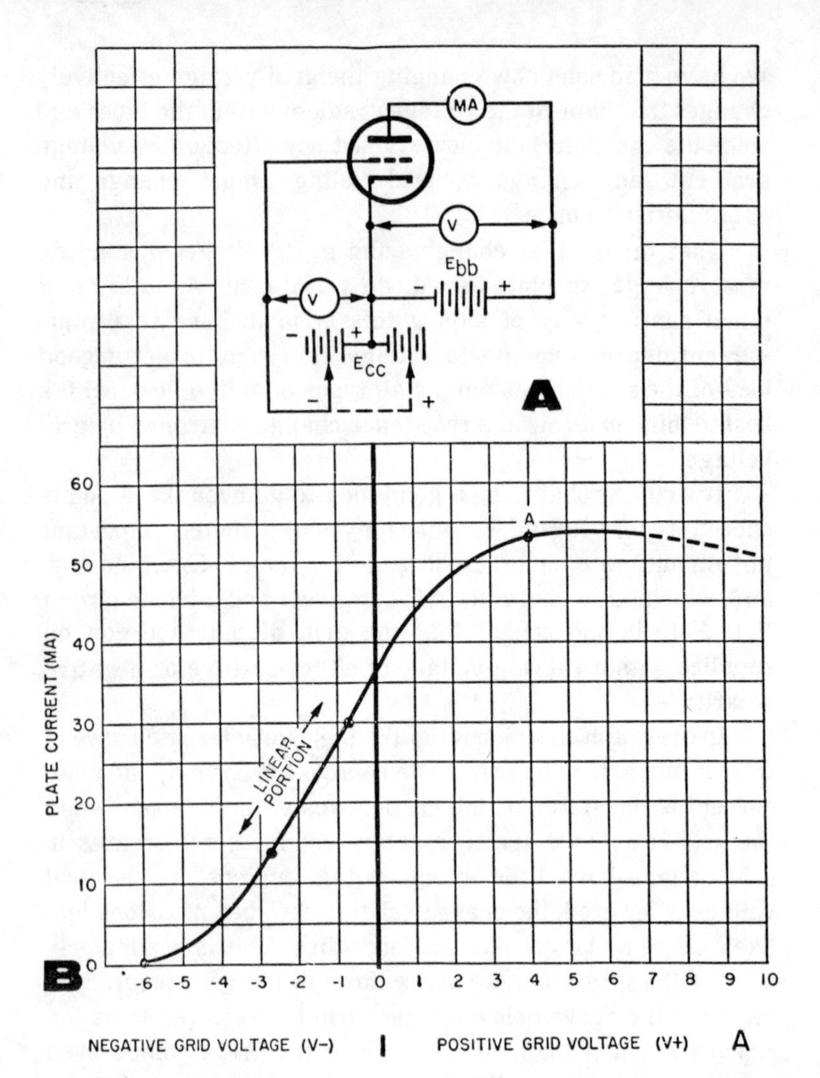

Fig. 5-15. When the grid voltage is varied from a positive voltage (saturation) to a very negative voltage (cutoff), the plate current varies between its maximum value and zero. If a load is connected in series with the plate of this tube—say in place of the milliammeter there—it will be "turned on and off" at the same rate the grid voltage is varied. Note the striking similarity to the relay's principle of operation.

resistance. Now in the vacuum tube we have a device which offers continuous control of resistance between wide limits.

In the vacuum tube we can achieve relatively low resistance by saturating the tube and causing maximum plate current to flow, or very high resistance by "cutting off" plate current. In neither case do we reach the resistance extremes available in the relay; but on the other hand, the relay does not give us the continuous control between these limits.

Where the controlling factor in the relay was power in the coil, the controlling factor in the triode is grid voltage. As grid voltage varies from **cutoff** to **saturation**, plate current varies from minimum to maximum.

The term **cutoff**, incidentally, means that value of negative grid voltage at which making the grid more negative causes no more reduction in plate current. It's the lower limit of control. Originally it was thought that plate current was completely cut off, but sensitive instruments show that a few microamperes of current usually do continue to flow. Most tubes are considered to be cut off when plate current is reduced to 0.01 mA.

One of the differences between a relay and a triode is that the relay acts as a switch while the triode is not dissimilar to a variable resistor. Another difference is the response time. It takes an excellent relay to operate within a thousandth of a second. The run-of-the-mill triode tube can respond faithfully to signals which go through a complete cycle in a mere hundred-millionth of a second. With a little extra care in construction, triodes can switch at rates up to 1000 MHz, where a signal cycle lasts only a billionth of a second.

Response time of a vacuum tube is determined primarily by the physical distance between cathode and plate. It takes electrons a definite, measurable time to make this trip, and the tube can operate only when this **transit time** is much smaller than any significant part of the signal cycle. Thus, to operate well at 100 MHz, a tube's transit time must be less than a billionth of a second. To operate at 1000 MHz, transit time must be less than a ten-billionth of a second. This is about the upper limit for conventional tubes, since the spacing from cathode to plate must still be large enough to get a grid in between and leave a little space for insulation.

Now that we have established that a triode operates by serving as a resistor whose value is controlled by the voltage between grid and cathode, let's look at a few simple circuits including the triode and see what happens to signals going through them. We're not going into great detail about the circuits; that comes in Chapter 6. The points we're going to be

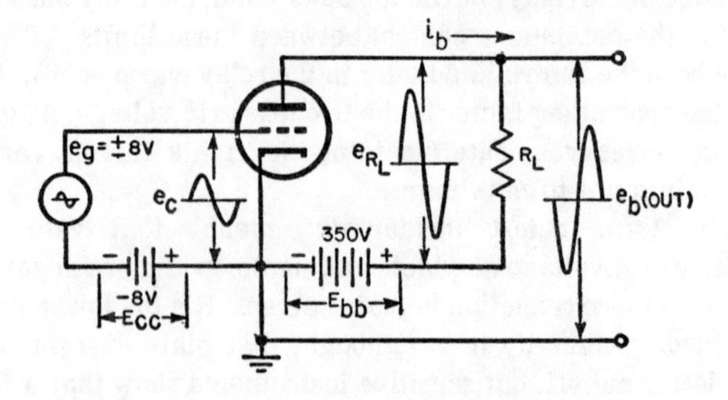

Fig. 5-16. Most common vacuum-tube circuit is this grounded-cathode version. If you consider the plate-to-cathode space as a resistance, you can see that the circuit is that of a voltage divider, with the load resistor (RL) as the upper leg. The circled sine wave represents an ac voltage that varies between plus and minus 2V. Depending on the polarity of the ac signal at any instant, it will either **aid** or **buck** the battery's potential of 8V. The grid will then "see" a voltage that varies between 6 and 10V. Negative-going input signal drives tube resistance up, raising plate voltage because of voltage-divider action. Result is that circuit's output signal is mirror image of the input signal, but 180 degrees reversed in phase. If two such stages are cascaded, each will reverse phase, and second-stage output will be in phase with input to first stage.

looking at here are primarily those involving the tube itself and how it can be used.

One of the most straightforward circuits involving a triode is the grounded-cathode arrangement shown in Fig. 5-16. Here the incoming signal is applied between grid and cathode to control the tube's cathode-to-plate resistance, and the plate circuit contains a series resistor which drops approximately half the supply voltage in the normal condition.

If the supply voltage is 350V as shown, and half drops across the resistor, then the tube's resistance and the load resistance must be equal and the plate voltage will be 50.

If we decrease grid voltage slightly, with an incoming signal, the tube's resistance will rise. This will increase the total resistance in the circuit, reducing current, and as a result less voltage will be dropped by the load resistance. The voltage at the plate, then, will rise.

If we increase grid voltage slightly, so that the tube's resistance decreases, total circuit current will increase. This will cause more voltage to be dropped by the load resistance, and the voltage between plate and cathode will decrease.

Thus, an increase of grid voltage causes a corresponding decrease of plate voltage in this circuit, and vice versa. We describe this state of affairs by saying that a "180-degree-phase shift" occurs between grid voltage and plate voltage, or that the grounded-cathode circuit "reverses phase" of the signal.

Note that the phase inversion is due to the change in resistance together with the fact that the resistance of the tube forms the lower leg of a voltage divider; it is not inherent in the tube's construction.

In fact, if we change things around just a little bit to form a grounded-grid circuit as shown in Fig. 5-17, we will find the phase reversal vanishing. The only difference between the grounded-grid circuit and the grounded-cathode circuit is that in Fig. 5-16 both the plate and grid circuits return to the cathode, while in Fig. 5-17 the plate circuit returns to the grid rather than to the cathode.

However, this one simple change actually reverses the roles of grid and cathode so far as the input signal is concerned. Now when the input signal rises, the cathode goes positive with respect to the grid—which is the same thing as the grid going negative with respect to the cathode, and increases the tube's plate-to-cathode resistance. Thus, plate voltage rises too, and the phase reversal does not occur. Output voltage is in phase with input voltage.

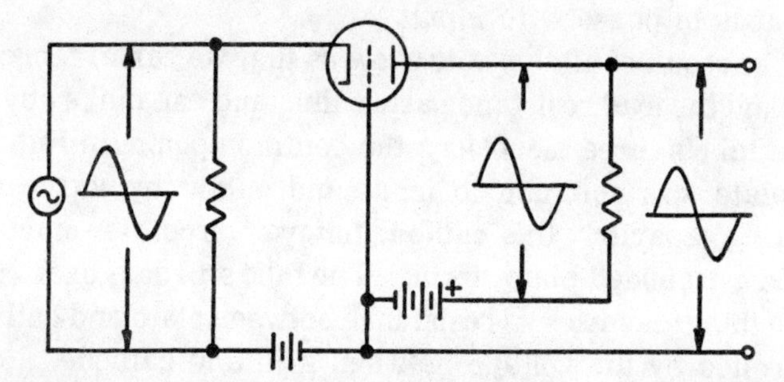

Fig. 5-17. Reversing roles of grid and cathode produces the **grounded-grid** circuit. Resistor from grid to cathode is normally very small and does not appreciably affect circuit action, but is necessary to provide complete dc path through plate and cathode. If signal source has continuous path for dc, this resistor is not needed. Reversal of grid and cathode roles eliminates phase reversal since driving cathode positive (with respect to grid) has same effect as driving grid negative (with respect to cathode). In either case tube resistance rises and more voltage is developed at plate, so output of grounded-grid circuit is in phase with input signal.

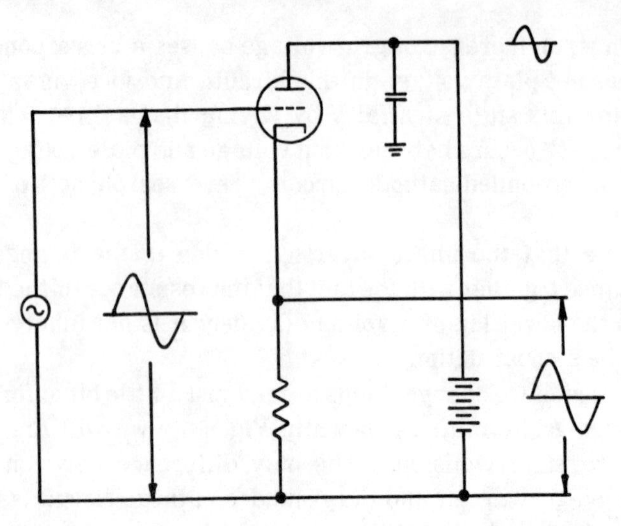

Fig. 5-18. This circuit moves the tube to the upper leg of the voltage divider. Now, when tube resistance becomes lower as a result of grid going more positive, reduced resistance of upper leg causes voltage across lower leg to rise, so output voltage rises in step with input voltage. Circuit is known as a **cathode follower** because output, taken from cathode, faithfully follows input signal.

Similarly, we could move the tube to the upper leg of the voltage divider, as shown in Fig. 5-18. Again, the only difference between Fig. 5-16 and Fig. 5-18 is the position of the load resistor. When an increase of grid voltage because of a rise of input signal level occurs, tube resistance becomes lower, and the voltage across the load resistor now rises. Output is in phase with input.

The result of all this is to show us that we can arrange the tube and the load resistance as we like, and can make any one of the tube's three electrodes the common point (in Fig. 5-18 the plate was common to input and output by virtue of a shorting capacitor; this **cathode follower** circuit is sometimes called a grounded-plate circuit). The tube still does exactly the same thing: it varies its resistance between plate and cathode, controlled by the voltage between grid and cathode.

Actually, the three circuits shown in Fig. 5-16 through 5-18 have radically different **external** characteristics, but the differences are not directly due to the differences of ground point. They result instead from effects of feedback, which we'll get into in the next chapter. This point is important, because when transistors first appeared many myths arose about their characteristics which were actually effects of the circuits in which they were used, and not of the fact that they were transistors.

That's a little ahead of us for now, though. Before we delve into the workings of transistors, there's a final corner of the vacuum tube's universe which we must take a look at. That's the area dealing with **multigrid** tubes.

Extending the Action—Multigrid Tubes

For a triode to operate properly, the action should be only one way. That is, the grid voltage should control tube resistance, but the effect of changing the tube's resistance should not come back to cause any additional change of grid voltage.

At low frequencies, triodes work this way, and all is well. At higher frequencies, the inherent capacitance which exists between grid and plate (Fig. 5-19) becomes an impedance small enough to permit energy to feed back from output circuit to input circuit. Since any capacitor introduces some phase shift in addition to the 180-degree reversal produced by the grounded-cathode circuit, this energy may either be **aiding** or **bucking** with respect to the original signal.

If we're attempting to use the triode to amplify radio-frequency signals, we will be using tuned tank circuits as load impedances rather than resistances. If you examine the three circuits of Fig. 5-20 carefully, you'll see that there are no really significant differences between **rf** circuits and those

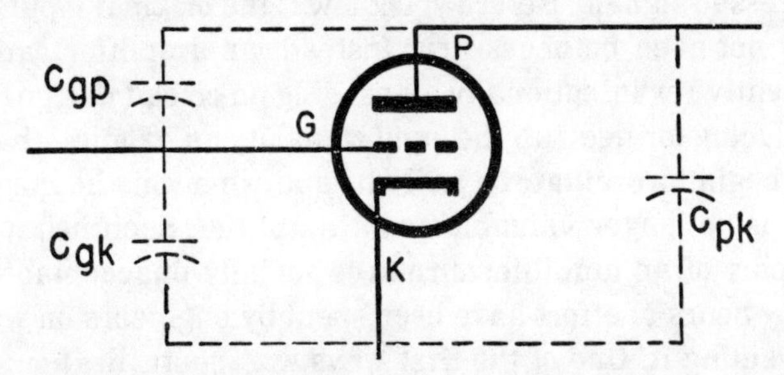

Fig. 5-19. The broken lines in this sketch show the effective capacitances that exist between the electrodes of all triodes. A capacitor, remember, exists when a pair of metal plates (tube electrodes) is separated by an insulator (evacuated space between electrodes). Of most concern is the grid-to-plate capacitance (C_{gp}). While too small to be significant in low-frequency circuits, it can become a real headache to those who want to use triodes at radio frequencies.

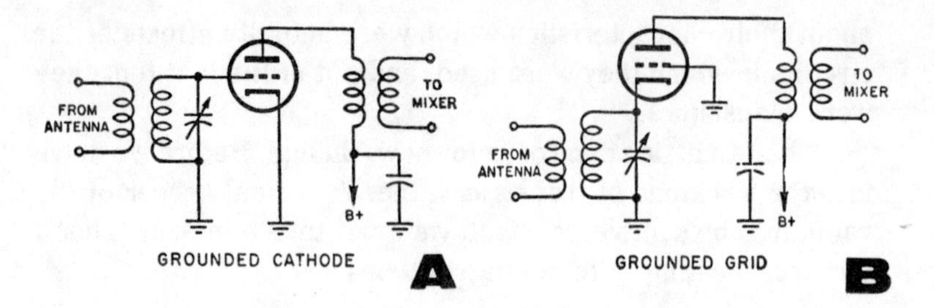

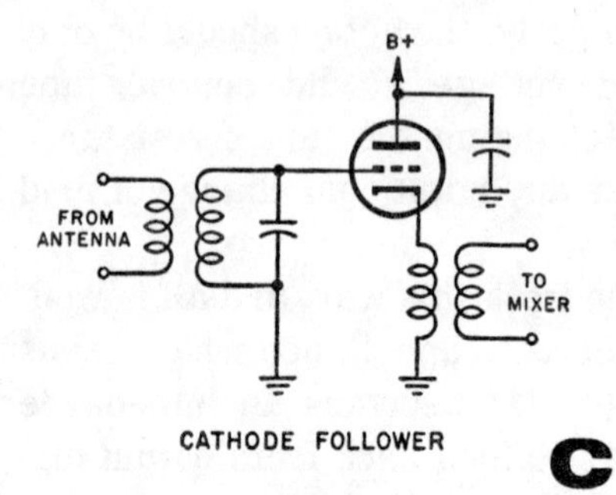

Fig. 5-20. When tuned circuits are used rather than resistors, the result is an rf amplifier. The circuits shown here are the radio-frequency equivalents to the conventional grounded-cathode, grounded-grid, and cathode follower audio-amplifier circuits we've just seen. In rf circuits, the internal capacitances between the elements of the tube itself can cause coupling even when it's not wanted.

we've been discussing. (The circuits shown are basic rf amplifiers used in radio receivers.) When tank circuits enter the picture, new considerations are required; rf implies very high-speed switching on the part of the tube involved. And when a tube operates very fast, it becomes easy for inadvertent signal coupling to take place—even through the inherent capacitances of the tube itself. When unwanted coupling occurs, strange things can happen to the phase of processed signals. Before you know it the original input signal may not even be necessary. Instead, an amplifier can inadvertently (or intentionally, depending on circuit design) begin producing for itself all the input signal it can handle—that is, it can begin to **oscillate** on its own; and when this happens, the tube is no longer valuable as an **amplifier.** Such behavior on the part of an amplifier circuit is socially unacceptable, and many hours of effort have been spent by engineers on ways of preventing it. One of the first ways was **neutralization**, which cancels out the feedback through grid-plate capacitance by using other feedback deliberately supplied in opposite phase. Another was the grounded-grid circuit, in which the grid is made common to both input and output and so cannot provide a feedback path.

But the one which turned out to be most profitable was the invention of the multigrid tube.

It started with the tetrode. To a conventional triode, a second grid was added between control grid and plate. This grid was grounded so far as signals were concerned, and so served to screen the control grid from the plate. From this function came its name of screen grid, or just **screen.**

While the screen is grounded for signals, by means of a bypass capacitor, it usually carries a dc bias voltage. This introduces another kink in the voltage-gradient curves, and makes it possible to achieve much greater gain with a multigrid tube than with a triode.

In effect, the screen voltage sets the operating point of a tetrode tube, while the control grid permits a signal to be introduced. Because the screen's voltage remains constant regardless of signal, the tube's characteristic curves have a much flatter slope, which means that the tube has high ac plate resistance.

Unfortunately, the tetrode introduced a new problem. The positive voltage on the screen grid gave electrons an additional kick en route to the plate; thus, it became possible to accelerate electrons to such velocity that they caused more electrons to fly free of the plate, when they splashed into it, resulting in a **net loss of electrons at the plate**! This **secondary emission** could result in electrons flowing from plate back to screen; and under certain circumstances, they did just that.

When these conditions were set up, the result was that the higher you made plate voltage, the less plate current you had (it was going to the screen instead). That works out to be a **"negative" resistance**; and if you happened to have a tuned circuit around, this negative resistance could cancel out all the real positive resistance in it. The circuit would then, free of all its resistance, oscillate merrily. However, the idea behind invention of the screen-grid tube was to stop oscillations, not give them a new way to happen.

Lest you believe this was only an exercise in imagination, note that an oscillator circuit called the **dynatron** circuit made use of the negative resistance effect in tetrodes, as long as any suitable tetrode tubes could be found on the market.

Even when you didn't quite get the conditions necessary for oscillation, the kinks in the curves of tetrodes put there by secondary emission caused headaches galore.

So what happened? They put still another grid in to suppress the secondary emission and called it, logically, the **suppressor** grid. This step brings us to the pentode or five-element tube. The pentode was the heart of most radio equipment in the years 1933 through 1965. Small pentodes were used as amplifiers in rf and intermediate-frequency (i-f) radio and TV receivers. (Unless your TV is solid state, it has several of these versatile little tubes.) **Power** pentodes were used in transmitters.

At very high power levels, such as those necessary in output stages of audio amplifiers or for radio transmitters, the suppressor grid of the pentode introduced some problems of its own. They were solved by using some beam-forming plates connected to the cathode which routed some of the space charge into the area between screen and plate, to form a **virtual cathode** where the suppressor of a pentode would be. This design, known as a **beam power tube**, has provided the mainstay of high power amplification from its introduction about 1939 right up until very recently.

As vacuum tube design progressed, many special features were added. These ranged from **variable-mu** or "remote cutoff" tubes, in which the grid wires were not spaced uniformly in order to permit an extended range of operating points, through special **converter** tubes having as many as five grids to combine the actions of local oscillator and mixer in radio receivers, to multiple tubes which contained two, three, or more independent sets of elements in the same envelope.

Tubes were made of glass with plastic bases, then metal, and finally of all-glass with metal leads which formed the pins. Special-purpose tubes have been made of ceramic. In size, tubes range from subminiature units the size of a pencil eraser to giant "jugs" as tall as a man.

In addition to the vacuum tubes used as active elements, another group made use of the emitted electron beam as a visual indicator. The beam strikes a phosphor, causing it to glow. Since the beam can be deflected by plates within the tube, and its strength controlled by grid action, the display can be manipulated as desired. One such tube is the oscilloscope's cathode-ray tube, which evolved into the picture tube of television. Another was the "magic-eye" tube used as a tuning indicator for many years. The major difference between these indicator tubes and the more conventional type is simply that the indicators produce no electrical output signals; their visual display is their reason for existence.

The vacuum tube is far from dead, but in many applications it is being displaced by the magic of solid-state electronics. The cornerstone of solid-state activity is the bipolar transistor, which we'll examine next.

TRANSISTORS AND SUCH

Just 42 years after Lee deForest put the grid in Fleming's valve and launched the era of the vacuum tube, a team of physicists at Bell Telephone Laboratories managed to make a crystal amplify, without benefit of vacuum, thus winning themselves a Nobel prize and establishing the age of solid-state electronics.

In the years since that day in 1948, the transistor has revolutionized electronics. Its many offspring now threaten to make the comic-strip "wristwatch two-way TV" a possible reality, and entire computers have already been built in a space smaller than a tray of ice cubes.

Today, any practical applications of electronics must involve transistors and their relatives in some way, since they are now the most commonly used of all active devices. So let's see how they operate...

We have just seen that vacuum tubes serve as active circuit elements in the form of resistances, the value of which can be varied by changing the voltage between grid and cathode. The transistor is similar; its very name comes from "**transfer**" and "**resistor**," and it also serves as a variable resistance. The controlling factor, however, is current flow in the base-emitter junction rather than voltage between grid and cathode. To see how it works, we'll have to back up all the way to primary physics again, but it shouldn't be too difficult.

Basic Semiconductor Principles

Way back in Chapter 1, we found that according to modern physics, all substances fall into one of three classes—insulator, conductor, and semiconductor.

We also saw that the factor which determined exactly which of these classes any specific substance occupied was the spacing of the **energy bands** in its molecular makeup. Each energy band can hold exactly two electrons at any time. If the energy bands of a substance are exactly filled, that substance cannot accept or donate any electrons outside itself, and so is an insulator. If there is only one electron per band, or if the bands are so closely spaced that no gap exists between a filled band and an empty one, electrons have room to move through the material and it is a conductor.

Finally, the case which interests us now, if small gaps exist between filled energy bands and vacant ones, so that the material acts like an insulator at low temperatures but tends to become a conductor when the temperature rises and drives the bands closer together, the substance is a semiconductor.

The way in which the resistance of a material changes when heat is applied can be used to decide whether it is a conductor or a semiconductor. The resistance of a semiconductor falls dramatically as temperature rises. That of a conductor, on the other hand, shows a moderate increase. Resistance of an insulator also falls as temperature rises, but not so markedly as that of a semiconductor.

To understand how semiconductor devices work, we're going to have to go into how crystals are held together, and while we do this we can see why the resistance of conductors rises with heat while that of semiconductors falls.

Any crystal is held together by the "sharing" of electrons between atoms. This is as true of insulating crystals such as quartz as it is of conducting crystals such as copper or aluminum. The atoms in the crystal are held firmly in position by atomic forces acting upon the electrons which occupy the energy bands. You can think of the two electrons in a single band as belonging to two separate atoms, but shared and swapped in such a manner that each electron makes a figure-8 orbit around both atoms and thus links them together.

The actual picture is much more complicated than that, because there may be many more than two atoms involved in each linkage, and the electrons may be more like shells surrounding the atomic nucleus than like planets orbiting a central body.

In a conducting material, any free electrons drifting through the structure under the influence of external electric forces always find a vacant energy band in their path, without regard to temperature, since a conductor is defined as a material which has such vacant bands available.

However, as temperature rises, the entire structure vibrates. Heat, actually, is simply the vibration of molecules. As it vibrates, the free electrons stand a greater chance of running into nuclei—or at least, having to spend some energy getting around them. This increased energy expenditure shows up as increased resistance.

In a semiconductor, on the other hand, availability of a path for free electrons to move along depends entirely upon temperature. At moderate temperatures, the path may not exist. As temperature rises, the path becomes available to high-energy electrons but not to those with less energy. This means that higher voltage is necessary to overcome the effect of the wide gaps between energy bands—and this also shows up as resistance. As the temperature continues to rise, the energy required to force electrons through the substance decreases, and resistance falls.

The semiconductors used in electronics are elements which have four valence electrons each. Because of the four valence bonds, these semiconductors form a basic crystal structure such as that shown schematically in Fig. 5-21. Each atom can be thought of as being at the center of a cube, and tied through its four valence bonds to four other atoms at

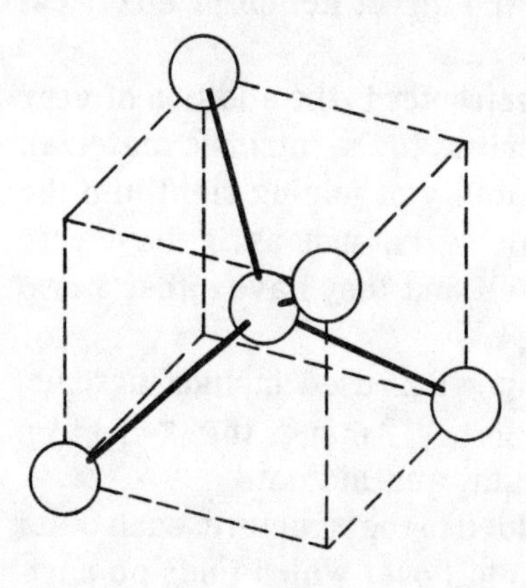

Fig. 5-21. Crystal structure of tetravalent elements such as germanium and silicon follows pattern such as that shown here, in which each sphere represents one atom. Each atom's four valence electrons form valence-pair bonds with four other atoms, which are located at corners of imaginary cube surrounding central atom as shown. Each corner atom is also the center atom of another such group of five, and the center atom shown here is a corner atom in each of these four additional groups.

corners of the cube which permit maximum distance between each atom and its neighbors.

A number of semiconductors fit this structure. One is carbon, which has two crystal forms. One, graphite, is soft and has energy gaps so small that it is often considered to be a conductor. The other, diamond, is the hardest substance known, and its energy gaps are so great that it is thought of as an insulator.

The next on the list as we go up the sequence of increasing atomic weight is silicon, one of the major ingredients of sand and one of the most common elements on earth. After silicon we find germanium, a rare element. Silicon and germanium are the most widely used semiconductors in electronics. The next semiconductor, tin, comes in two forms like carbon. Unlike carbon, both forms of tin have gaps so closely spaced that for all practical purposes they act as conductors except for their resistance when heated.

The effects which make possible solid-state electronics occur only in single crystals of material. A crystal of either silicon or germanium is composed of many millions of atoms, all bonded firmly to each other in the arrangement shown in Fig. 5-21. Under laboratory conditions, such a crystal can be grown to almost any size desired, and can be made almost perfectly pure.

A crystal of pure semiconductor material, called in the jargon of semiconductor physics **intrinsic** material, isn't much use, though. In its pure form, the semiconductor is essentially a very high valued resistor. Its resistance can be changed by the action of heat, but not by the direct action of electrical forces.

The secret of solid-state electronics is the addition of very small amounts of specific impurities to the intrinsic material. These impurities have the capability of joining right into the crystal structure at the atomic level, just as if they were atoms of the semiconductor itself, but they have either 3 or 5 valence electrons rather than 4.

Some of the 5-electron impurities used include arsenic, phosphorus, antimony, and boron. Among the 3-electron additives are aluminum, gallium, and indium.

If a 5-electron element is added to the structure, each of its atoms has one valence electron left over which finds no partner in the crystal structure. This produces a surplus of electrons, so these impurities are called **donor** substances because they donate extra electrons. Addition of donor substances to the mix provides an excess of negative charge, so the resulting crystal is called **n-type** material.

Similarly, adding a 3-electron element to the structure results in "holes" where surrounding semiconductor atoms have electrons, but the impurity has no matching electron to fill the offered valence bond. These "holes" will accept any wandering electrons that happen by, but the associated atom is then electrically unbalanced and so repels the electron. Undaunted, the hole then accepts the next passerby. The net effect is that the hole appears to wander throughout the structure. Such elements are called **acceptors**.

Crystals containing acceptor impurities have a shortage of negative charge, which is the same thing as saying positive charge, and so are called **p-type** material.

When a crystal consists entirely of either type, the excess electrons (if n-type) or holes (if p-type) tend to diffuse throughout the crystal structure. Thermal energy causes them to be in constant motion, but the direction of motion is random, so no current flow occurs.

When single crystals are being grown, it's possible to implant donor material at one point, then switch to acceptor elements, and so produce a single crystal which is composed of two different types of material. The surface at which the type changes from one to the other is known as a **junction**, and the key events of solid-state electronics happen at junctions.

At a junction, the surplus electrons of the n-type material tend to diffuse across the junction to fill the holes of the p-type. These "combinations" form valence bonds which can no longer diffuse throughout the crystal. However, the first few recombinations form an area known as the **depletion region** which is essentially free of either excess electrons or holes, and so forms an insulator as good as intrinsic material would be.

The diffusion of electrons and holes across the junction corresponds exactly to the emission of electrons from a heated cathode. Their combination to form a depletion region corresponds to the space charge which surrounds a heated cathode. The depletion region, acting as an insulator, forms a barrier to continued combination, just as the space charge always exactly balances out the energy of electron emission. Thus, we have a situation at the junction very much like that which exists in a diode vacuum tube with its cathode hot but no applied plate voltage.

In the vacuum diode, if we applied positive voltage to the plate it would attract electrons out of the space charge, thus making room for more electrons in the space charge region and permitting a current to flow. Negative voltage on the plate repelled electrons, blocking any flow of current.

In a semiconductor junction a similar situation occurs. If we apply positive voltage to the n-type side of the junction, it tends to draw electrons out of the n-type material. At the same time the electrons available at the negative side of the power supply fill the holes of the p-type region, so that the depletion region eventually extends throughout the entire structure of the crystal.

This situation prevents current flow, and is known as **reverse bias**. No current flows through a reverse-biased junction until the voltage becomes great enough to cause a breakdown of the crystal structure. (Actually, very small leakage currents may flow, but they are many times smaller than normal signal current in the circuit.)

If we reverse the connections, so that the negative terminal of the power source connects to the n-type side of the junction, electrons will be forced away from the power source toward the junction. At the same time, holes will be forced away from the power source, also toward the junction. Thus, the electrical pressure imposed by the power source overcomes the barrier and compresses the depletion region, until it eventually disappears. When the barrier is breached, current flows freely through the junction. In germanium, this occurs at about 200 mV (0.2V) pressure; with silicon, the barrier voltage is about 600 mV. When the barrier has been overcome, the junction is said to be in a state of **forward bias**.

Figure 5-22 shows a single pn junction in all three conditions. At A, it is shown in the absence of external connections. Note the barrier region width. At B, reverse-bias conditions are shown. The barrier is much wider. Shown at C is a forward bias condition just before the barrier is overcome, while D shows the action after full forward bias is established. The **plus** signs represent holes and the **minus** signs represent free electrons in these illustrations.

Semiconductor Diodes

Since the single junction behaves much like a vacuum diode, one would expect that it could be used in its place. In general, this is true.

Certain differences, however, may limit the idea of plug-in replacement. A vacuum diode, for instance, can withstand very high reverse voltages without damage, and even if it does arc over because of excess voltage may not be damaged. A semiconductor diode, on the other hand, may be destroyed by voltages normally considered to be low, and is almost certain to be ruined by any overvoltage condition.

Reverse resistance of a vacuum diode is so high as to be effectively an open circuit. That of a semiconductor diode may be as low as 5000 ohms, and few have reverse resistance higher than a few hundred megohms—which may not be enough in some circuits.

With these limitations in mind, however, the semiconductor diode can be used to replace the vacuum diode—and in most areas already has almost completely taken over.

Power supplies, for instance, once used vacuum diodes to convert ac to dc. Now, silicon junctions do the job, in a very tiny fraction of the space and with considerably less power loss. The vacuum diode lost up to 15V across the resistance of the tube. The silicon junction loses only the 600 mV required to overcome the barrier. The vacuum diode required 5 watts or more power to heat the filament. The silicon junction runs cool. Because of the filament voltage requirement, most vacuum diodes used only a centertap full-wave rectifier circuit which required twice as much copper on the transformer as the circuit being supplied actually needed. The silicon junction can be used in a full-wave bridge circuit with only half as large a transformer. The list of advantages goes on and on.

Similarly, both audio and video detection were once done by vacuum diodes. Germanium junctions now do the job in much less space. Here, the advantages are higher frequency

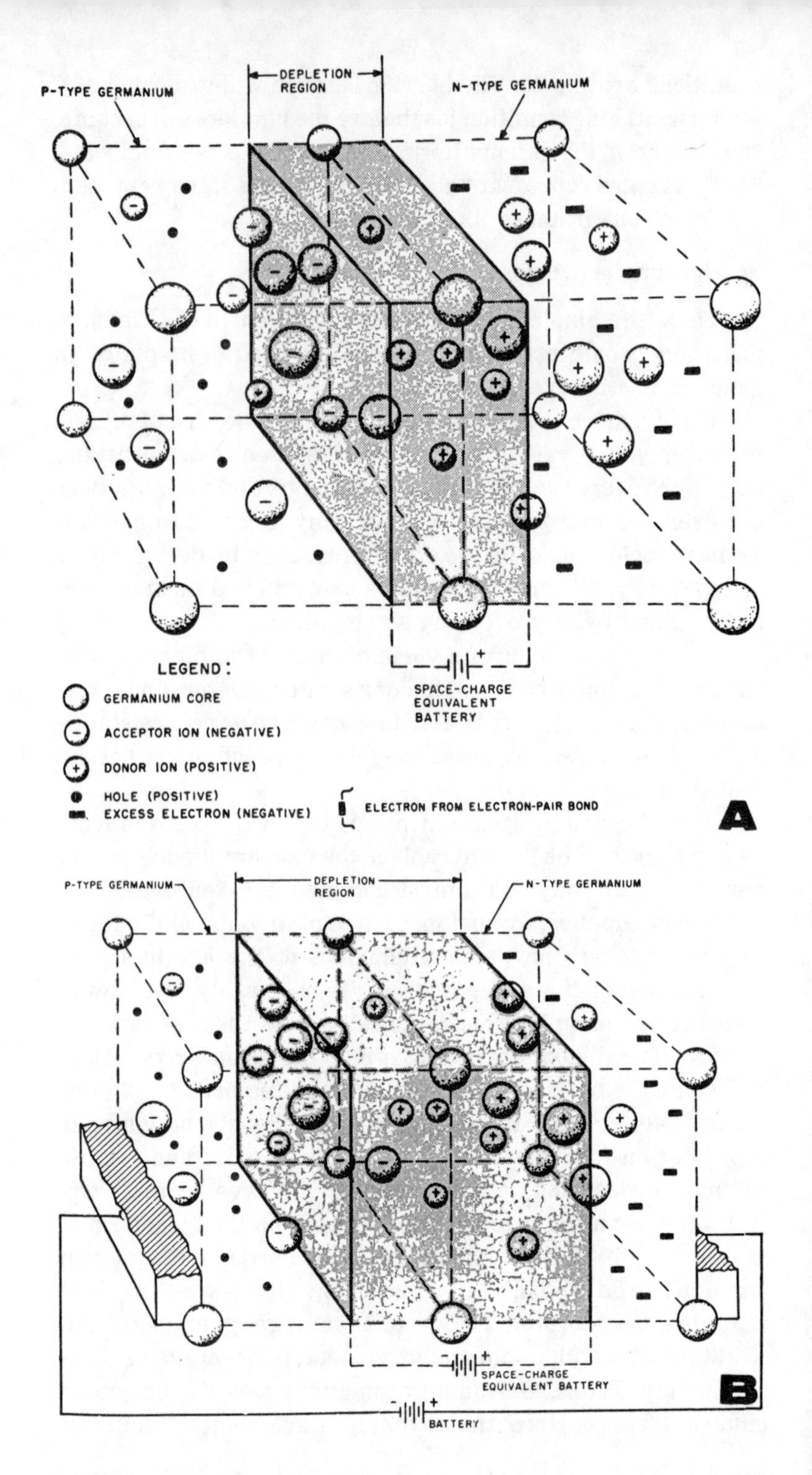
DEPLETION REGION
P-TYPE GERMANIUM
N-TYPE GERMANIUM
SPACE-CHARGE EQUIVALENT BATTERY
LEGEND:
GERMANIUM CORE
ACCEPTOR ION (NEGATIVE)
DONOR ION (POSITIVE)
HOLE (POSITIVE)
EXCESS ELECTRON (NEGATIVE)
ELECTRON FROM ELECTRON-PAIR BOND
A
P-TYPE GERMANIUM
DEPLETION REGION
N-TYPE GERMANIUM
SPACE-CHARGE EQUIVALENT BATTERY
BATTERY
B

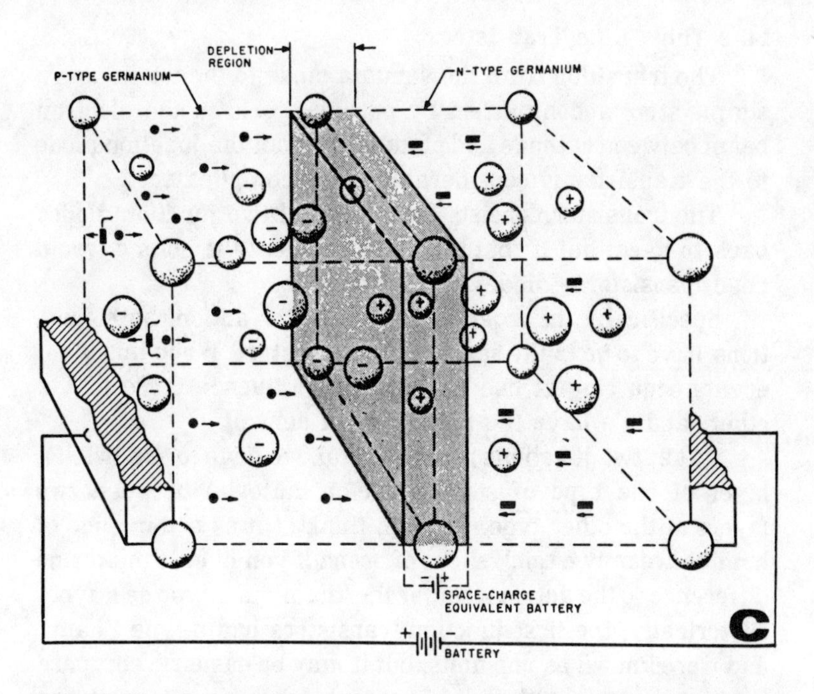

Fig. 5-22. Single pn junction is shown in absence of bias voltage (A), with depletion region established through normal recombination of carriers in each type of material with reverse bias (B), which increases width of depletion region by forcing carriers to ends of material; and with forward bias not quite adequate to establish conduction (C), which forces carriers toward junction and so reduces width of barrier.

response, less stray capacitance, and, as always, the lack of any need for filament heating power.

One application for the junction diode which would work with vacuum tubes was hardly ever done with them, because the semiconductor diode was available when the circuitry became essential. That's in logic switching, such as is presently done in computers. The diode's forward and reverse bias conditions are similar to the open- and closed-contact conditions of a relay, and can be used for switching. Specifically, a diode can be switched from forward to reverse bias by application of voltage alone. Very little current need flow through the junction to maintain the forward-bias condition. When forward bias is present, though, the circuit is just like a piece of wire for any signals which may be present so long as they do not overcome the bias. Thus audio signals may be switched at will by applying or removing dc bias voltages from a network of diodes.

Like Tube, Like Transistor

The transition from the vacuum diode to the triode was a simple step, accomplished by putting a grid in the electron beam between cathode and plate. That from the junction diode to the transistor is considerably more complicated.

The transistor consists essentially of two junction diodes back to back, but if that were all there is to it no one would need transistors—diodes alone would do.

Specifically, in order to get transistor action, both junctions **have to be in the same crystal structure**. When this is so, events occurring at one junction can influence those at the other, and we have the possibility of control.

To get two junctions in one crystal, we have to sandwich a layer of one type of semiconductor material between two layers of the other type. You can think of it as a thin slice of ham between two thick slices of bread if you like. It makes no difference to the action whether the "ham" is p-type or n-type. Historically, the first junction transistors had n-type "ham" and were known as pnp units, but it may be easier to compare transistor action with that of vacuum tubes if we look at npn devices, which consist of a thin layer of p-type material between two slabs of n-type.

One of the outer layers is called the **emitter**, since in the final circuit it will be emitting charge carriers. In an npn transistor, the charge carriers are electrons. In a pnp device, the charge carriers are holes. With vacuum tubes, we dealt only with electrons and didn't have to use such a term as carriers, but since transistors come in complementary polarities, here we must.

The other outer layer is known as the **collector** because it collects the emitted carriers. These names of emitter and collector would fit the vacuum tube's cathode and plate equally well, but time and custom have established the terms and it's a bit late to try to change them now.

The middle layer is known as the **base**, and this name is based on historical rather than logical reasons. The original transistor was a point-contact device, rather than being based on junctions, and its emitter and collector were "point contacts" to the semiconductor crystal which physically formed a base for the whole device. Not until much later were the similarities between transistor action and vacuum tube action fully understood, and by then the name "base" for the middle layer, like "cathode" and "plate" in vacuum tubes, was established by usage.

When the emitter of an npn transistor is connected to the negative terminal of a battery, and the collector to the positive terminal, we find that the two junctions inside the transistor are oppositely biased. The collector-base junction, with the collector's n-type material connected to the battery's positive terminal, is reverse biased. The base-emitter junction, with the emitter connected to the negative terminal, tends toward forward bias. The only thing preventing forward bias from existing is the reverse bias on the upper junction, which prevents current flow. If we provide a little current through an external path from battery positive to the base, forward bias is established.

In this forward-biased junction, the emitter provides the electrons which stream across the junction into the base. We might expect these electrons to move right on out through the base lead to maintain the forward-bias circuit, but we would be wrong.

Like everything else, charge carriers follow the path of least energy expenditure, and it takes energy to change direction. Most of them stream right on through the base into the collector, despite the reverse bias existing at the base-collector junction. This reverse bias, after all, is not opposing their direction of travel—to them, it's still a downhill plunge.

The transistor designer does all in his power to make it easy for this to happen. The base, for instance, is physically very thin—about the thickness of a single piece of recording tape. The emitter is heavily doped with its impurities to provide plenty of available carriers, but the base is doped lightly to make it more difficult for any carrier to linger there or change direction.

In practice, 95 percent or more of the emitter current passes right through the base into the collector, and only 5 percent or less is diverted through the base lead. The ratio by which this current divides, in fact, is the most common measure of a transistor's gain. In the early days, the percentage factor alone was used. It was called **alpha**, and ranged from 0.95 to 0.99. A transistor with an alpha of 0.99 would divert only 1 out of every 100 uA of emitter current through the base lead, and the other 99 would go to the collector.

As performance improved, a related measurement called **beta** took the place of alpha. The beta of a transistor is a measure of its current-amplification ability. Our previous example of an alpha of 0.99 could also be described as having a current amplification of 99 times, since for every microampere injected into the base, 99 uA would flow in the collector circuit. Its beta, then, would be 99.

Typical values of beta for modern units range from 100 to 500.

A transistor having a beta of 500 would have an alpha of about 0.998. Increasing alpha to 0.999 would correspond to doubling the beta to 1000; so at very high-gain figures, beta is a more convenient measure than alpha.

Either way, we're getting ahead of ourselves, because we have not yet found out how these actions are related to the transistor's resistance in a circuit.

Recall that the junction diode serves as a switch; if it's forward biased, it's just like so much wire to a circuit. Thus, so long as we keep the base-emitter junction in forward bias, it cannot affect the resistance of the circuit between collector and emitter.

The collector-base junction, however, is a different matter. This junction is in reverse bias, and so has very high resistance. Recall one of the major differences between a junction diode and its vacuum counterpart; the vacuum diode's reverse resistance is nearly infinite while that of the junction in reverse bias is high, but finite.

The carriers plunging through the base from the emitter into the collector have to cause some action there; they drive an equal number of carriers out at the collector lead, so we do have a current flow through the emitter-collector circuit and this establishes a resistance.

Now we can look at the situation from at least two viewpoints. One way of putting it is that the carriers which plunge through the base tend to reduce the reverse-bias at the collector-base junction, and thus reduce its high resistance.

A simpler and more direct explanation is that the percentage of carriers which divides between base and collector is established by design of the specific transistor involved, and so any increase of current in the base circuit must produce a proportional increase of current in the collector circuit.

Like any oversimplifications, neither of these two viewpoints is precisely accurate, but either is close enough to the realities involved for all practical purposes. The concept of reducing resistance at the collector-base junction is in accord with our previous presentation of all active devices as being devices having one or more electrical characteristics capable of change by control signals.

Note that the control signal for a transistor is in the form of current into (or out of) the base-emitter junction, where that for a vacuum tube is the voltage between grid and cathode. Since current is the dual of voltage, it is only to be expected that a transistor is the dual of a vacuum triode. When due note is taken of the key differences, they can be substituted one for the other by making the circuit adjustments implied by duality.

We have already noted that the emitter of a transistor corresponds to the cathode of a tube, and the collector to the plate. It's almost unnecessary to add that the base of the transistor corresponds to the grid of the triode.

While multibase transistors have been built, they are not in current use. Circuit designs have gone in other directions, and the differences in mechanics between the transistor and the tube make solid-state versions of multigrid tubes unnecessary.

While we looked only at the npn transistor, the same explanation would hold true for a pnp by reversing all polarities. This is one of the major differences between transistors and tubes. Having both polarities available makes possible with transistors some circuits which could never be built using vacuum tubes.

During the first 20 years or so of the transistor's life, the name "transistor" was adequate to identify it. Then still newer types of semiconductors began to appear, and a distinctive name became necessary. As a result, the familiar transistor—the type we have just examined—is now known as a **bipolar** transistor to distinguish it from its younger relatives.

The name bipolar comes from the requirement for two polarities of power source, to provide forward and reverse bias on the two junctions. In practice, a single power source usually suffices, but the requirement that one junction be forward biased while the other is reverse-biased remains, and that's the identifying characteristic of a bipolar transistor.

OTHER SEMICONDUCTORS

Since the transistor opened the door to solid-state active devices, it has become almost impossible for anyone to keep up with all of them. Some have enjoyed a brief flurry of notoriety and vanished into obscurity, while others are rapidly becoming vital parts of the electronics industry. In this section, we can only skim the surface, omitting mention of more specialized semiconductor devices than we examine.

Most of the devices we'll look at here follow the same basic principles of operation as the transistor, although details will vary. All make use of the effects of impurities in semiconductor crystals, and are based on the injection and collection characteristics which make the transistor work. However, not all of them produce a variable resistance in a circuit. At least one important class of devices produces variable reactance, rather than resistance—but that's getting ahead of ourselves.

The Field-Effect Transistor

The conventional bipolar transistor has one major point of difference in comparison to a vacuum tube, which caused many headaches to circuit designers. Since a tube is operated by voltage rather than current, it takes virtually no power from its signal source. This means that a circuit using a tube does not impose any effective load upon the preceding circuit.

The bipolar transistor, on the other hand, is current controlled, which means that it must take power from the preceding circuit. Thus a transistorized circuit loaded the previous stage to a greater degree than a similar tube circuit would.

By clever circuit design, engineers found ways to get around the problem and reduce the loading to an acceptable degree. However, they still felt it would be nice to have some circuit element which combined the size advantage of the bipolar transistor with the voltage-controlled characteristic of the vacuum tube.

The answer to this wish is the field-effect transistor, or FET. Sometimes the type of construction is combined with the initials, making MOSFET or JFET, but they all serve approximately the same purpose and operate in essentially the same way.

Unlike the bipolar transistor, the FET has only one junction. It's a most unusual junction, though. The body of the FET (originally called a unipolar transistor) usually consists of n-type material, and the junction of p-type material forms a ring all around the body. Contacts are made to each end of the n-type material, and to the p-type. The junction is operated in a state of reverse bias, so there is always a barrier at the junction.

Figure 5-23 shows schematically what happens. When a voltage is applied across the n-type material, current flows through it. The reverse bias on the p-type **gate** increases the size of the depletion region at the junction, thus making a narrower path for the current and so increasing resistance between **source** and **drain** (same as bipolar transistor's **emitter** and **collector** terminals).

If drain voltage is increased, the current increases only moderately. The reason is that an increase of drain voltage

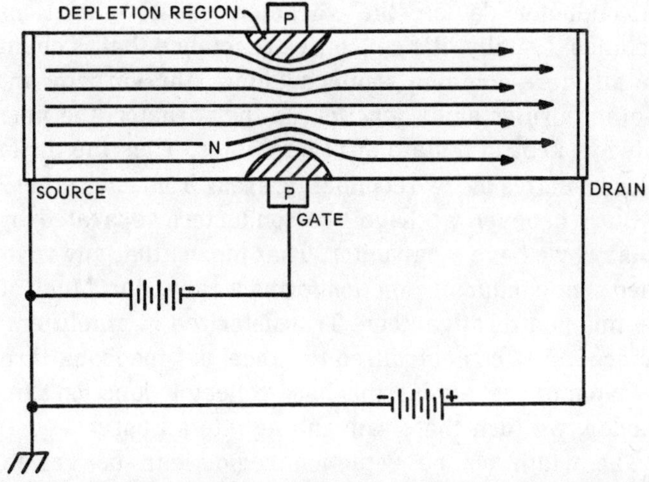

Fig. 5-23. Field-effect transistor (FET) works like vacuum tube in that **gate** is reverse-biased and draws almost no current. Current flow between source and drain is controlled by amount of reverse bias between gate and source, because size of depletion region determines the effective conducting cross section beneath the gate and so determines the resistance between source and drain. The greater the reverse bias, the wider the depletion region and so the higher will be the resistance. When depletion region extends completely across the n-type material, **pinchoff** occurs and source-drain current is cut off.

produces, by voltage-divider action within the FET body, a greater reverse bias, which "**pinches off**" the current path, so that some of the increased current flow due to higher voltage is restricted by greater reverse bias.

This causes the characteristic curve of the FET to look very much like that of a pentode tube. Above critical voltages, drain current is almost independent of drain voltage and is controlled entirely by gate voltage. This means that the ac plate resistance of the device is extremely high, which in turn means that great gain can be achieved in a single stage.

Since the gate junction is always reverse-biased, its resistance is similar to that of a triode's grid. Thus, the FET is currently the closest thing available to a "solid-state tube." The only thing it can't do is handle large amounts of power. But that may be only a matter of time.

Varactors

All of the active devices which we have looked at so far have changed the resistance in a circuit by means of a control voltage or current. However, resistance is not the only circuit characteristic which can be controlled. One class of semiconductor device, the **varactor**, causes variations of reactance. Usually, it's capacitive reactance that is changed. Like all more common semiconductors, the varactor makes use of impurities and a junction. In the varactor, the junction is always kept in a state of reverse bias. Thus the depletion region insulates the two conducting areas from each other.

But whenever we have two conductors separated by an insulator, we have a capacitor. That means that any reverse-biased semiconductor junction forms a capacitor. Much of the time this is a disadvantage. Transistorized rf amplifiers, for instance, must be neutralized to cancel out feedback through this capacitance across the base-collector junction. In the varactor, we turn the disadvantage into an asset.

The width of the depletion region can be varied by changing the reverse-bias voltage, as we saw earlier, and this has the effect of changing the distance between plates in our capacitor. As in the mica compression trimmer capacitor, where capacitance value is adjusted by varying plate spacing, this change in distance between the plates changes the capacitance.

Such a varactor is sometimes called a **voltage-variable capacitor**, which adequately describes its action. The major difference between a varactor and a garden-variety junction diode is that the varactor is specially designed to obtain the desired result at the expense of all other characteristics, while the everyday diode is intended for other purposes. Many of them will, however, sèrve admirably as varactors.

Since the junction is always reverse-biased, a varactor is voltage operated (like a vacuum tube) and draws virtually no power from its bias source. This is a marked difference from normal transistor operation.

Integrated Circuits

The size range between the largest and the smallest vacuum tubes is about 100 to 1; even if we exclude giant transmitting tubes and microscopic hearing-aid designs, the size range between say a 27-inch TV picture tube and the smallest tubes normally used in household electronics is nearly 50 to 1.

The first transistors, however, took only about 0.1 percent the space of the vacuum tubes they replaced, because the action of a transistor occurs at the atomic level inside a single crystal, while that of a vacuum tube takes place in an artifically maintained vacuum over a rather appreciable distance.

The reduction in physical size caused by the emergence of solid-state circuits was so dramatic that several years were required for most electronics enthusiasts to realize that, tiny as a transistor is, something like 99 percent of its volume is superfluous. The really active part of a transistor is microscopic in size; the rest is just to make it big enough to wire into a circuit.

By the time this became common knowledge, research had moved forward into the era of **integrated circuits**, which may have anywhere from 2 to 2000 or more transistors on a single transistor-sized wafer, or **chip**. While a transistor is just one active element, and must be connected into a circuit to accomplish anything, an integrated circuit (IC) usually is a complete circuit in itself, built into a transistor case but including all the rest of the circuit elements as well, and requiring only power signal connections.

The advent of the IC is slowly but surely bringing to an end the era of home experimenters laboriously assembling breadboard circuits, just as the advent of ready-made coils and transformers put an end to the do-it-yourself coil winding that characterized electronics experimentation in the 1920s and 30s. When a high-performance amplifier can be purchased for less than two dollars, why spend several days and ten times as much cash for an inferior substitute?

The great advantage of the IC is that it brings all the benefits (and yes, all the flaws, too) of mass production to the area of electronic circuits. If it costs $15 to build a single amplifier, it's probably possible to build 10,000 identical to it for $15,000, or $1.50 each.

When mass production is achieved by the IC route, in which most of the processing is automated, the cost reduction achieved by quantity production is even more dramatic. A typical computer IC chip providing two independent amplifiers in one can, using 15 transistors, sold for $7.50 each in 1965. By 1970, the same circuit could be purchased for just under a dollar. In 1973, it dropped to 15 cents in production quantities; and individual units can be found on the surplus market for a quarter each. At the 1965 price it was a bargain; today, it's a steal.

The biggest drawback to mass production in electronics is that the circuits must either be very general, or very common. The operational amplifier, for instance, is a general-purpose circuit that can be used almost anywhere—but sometimes it's like using a 10-ton truck for a paperweight. The prevalence of computer circuits available as ICs is an example of commonality taken to extremes. Computer circuits are very limited; only a few basic types are necessary for the most complex system. This makes it practical to mass-produce each of the basic types of circuits, with the assurance that someone will need enough of any of them to make it worthwhile.

IC technology is growing so rapidly that anything other than the most general description would be obsolete before this could be printed. In general, though, ICs are grouped into three major classes called **small-scale integration** (SSI), **medium-scale integration** (MSI), and **large-scale integration** (LSI).

In SSI, only a few circuit elements are in each IC. For instance, the transistors necessary for a differential amplifier might be on one SSI chip, or a single operational amplifier might be another example.

Most ICs in use today fall into the MSI class, where several similar circuits are contained on the same chip. The dual amplifier mentioned previously is an MSI chip, as would be many of the multistage computer ICs employed in current home construction projects.

Until recently, LSI was seldom found, because the demand had not yet created a mass market to make production practical. But with the computer industry beginning to produce memory systems using LSI techniques, and the advances in consumer electronics (like 4-channel stereo), you're likely to encounter this type of IC in the future at the neighborhood parts emporium.

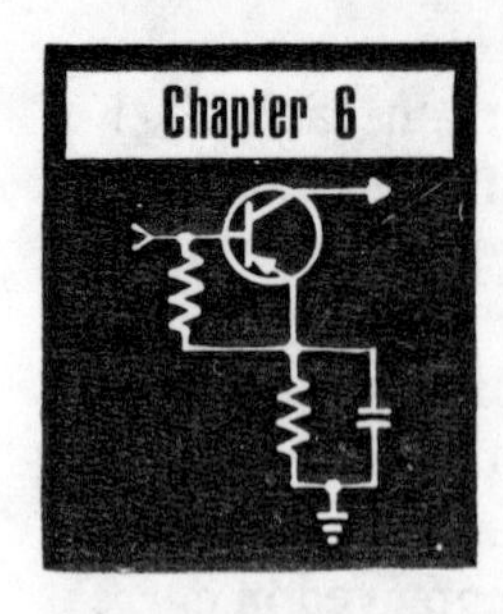

Useful Circuits

Back in Chapter 4, we made the acquaintance of simple circuits based on passive elements, and learned the basic principles of applied circuit theory. Passive elements, however, cannot be controlled automatically; for this, active circuit elements are necessary. In Chapter 5 we met the most common active devices and saw how they work. Now we're ready to combine active and passive elements, to yield useful circuits capable of controlling themselves to at least some degree.

Most useful circuits we'll meet in practice do combine active and passive elements. What's more, the vast majority of useful circuits are variations on two themes—**the amplifier** and **feedback**—usually together. A third theme, encountered less frequently, is that of the switching circuit.

In this chapter, we'll study all three of these basic themes found in useful circuits. With our knowledge thus consolidated, we will be able to apply it to problems we may encounter later.

First, we'll examine amplification, learning what it is and how it is achieved in practice. With a firm grasp on the principles which govern amplifiers, we will turn our attention to feedback, and find out why it plays such an important role in practical electronics. Finally, we'll look at the basic theory of switching circuits, upon which the entire computer industry and its offspring are based.

AMPLIFICATION

Just what is an amplifier, anyway?

Trying to answer that question in detail could take more words than this entire volume, most of which might have very little to do with practical electronics.

It can get complicated, because an amplifier need not necessarily contain either tubes or transistors, or, for that matter, even be electronic in nature.

Anything that increases effective power is an amplifier. One simple example of a mechanical amplifier is a burglar's pry-bar. The lever, in fact, was probably the first application of the principle of amplification.

But we're interested primarily in electronic amplification, so we need to limit the term "amplifier" a bit. At the same time, we want to keep the definition as broad as is practical, so that we don't limit ourselves unnecessarily.

When we make this compromise, we find that an electronic amplifier can be defined as "any circuit which increases the power level of an electrical signal." In turn, a **signal** in this definition must be pinned down as "a sequence of electrical power levels which, by their variation, carry some sort of information." This information may be the mere fact that the signal is present, as in the ac power "signal" at the wall outlet, or it may be as complicated as a composite color TV broadcast signal.

When we use these definitions, we will find that all the things that we might normally think of as "amplifiers" such as the key stages of TV and radio receiver, hi-fi sets, and the like, are **still** amplifiers by these definitions, while components such as transformers, which can increase the voltage **or** current level of a signal, but not its total power level (voltage **and** current), are rightfully excluded.

Most electronic amplifiers now in use are of a type known to engineers as **two-port** devices. This means simply that the amplifier circuit, if thought of as a black box, would require two ports, one to **enter** and one to **exit** from the box. The signal to be amplified is fed into the **input** port, and the amplified result is taken from the **output** port. In practice, we speak merely of the **input** and the **output** of the circuit.

Since any electronic circuit must be complete for power to move through it, each port requires two terminals. For this reason, some texts speak of two-port devices as four-terminal devices. However, almost all basic amplifier circuits are actually only three-terminal affairs, since the input port and the output port share one terminal in common.

In fact, the one common terminal between input and output port is a direct consequence of the basic amplification technique employed by all practical amplifier circuits—that of

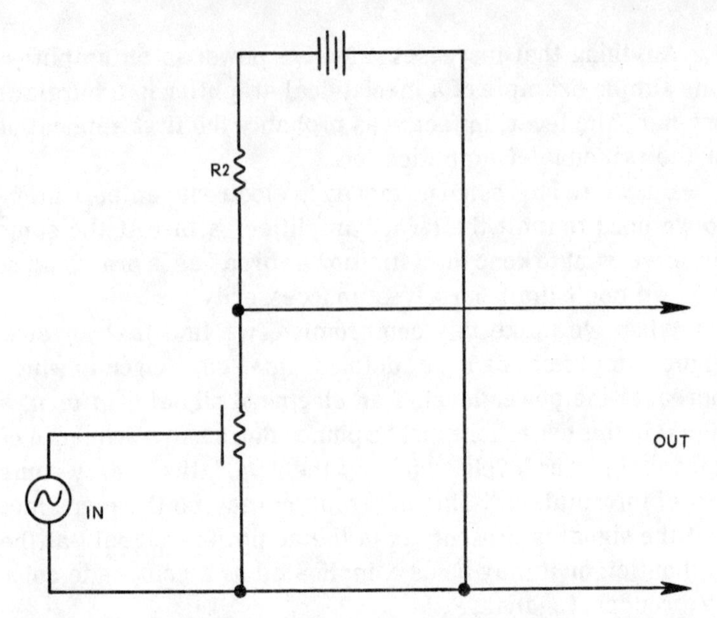

Fig. 6-1. This simple voltage-divider circuit shows the basic action of all amplifiers with which we'll be dealing. Lower resistor is actually an active device such as vacuum tube or transistor, the resistance of which is controlled by the input signal shown applied to bar alongside symbol. This controls current through series resistors, and changes the voltage at the output terminal. Power furnished to the output comes from the steady source represented by the battery, but is controlled by the input signal.

valving power from an external power supply through an active device, the resistance of which is controlled by the input signal.

We have already seen that both vacuum tubes and transistors appear in their circuits as resistors which can be varied in value by a control signal. In the amplifier circuits, this control signal is furnished by the input signal, and the change of resistance is made to cause corresponding power changes in the output signal, thus achieving the desired amplification.

Figure 6-1 shows this schematically. The bar beside resistor R1 is there to indicate that this is the control signal which changes R1's resistance, because R1 really is not a resistor at all. Instead, it's the active device in the amplifier circuit. R2 is a load resistor, the battery represents the power supply which furnishes steady power, and the circle with a sine wave in it (a generator symbol) represents the input signal source.

The similarity between this circuit and the voltage divider of Fig. 4-4 should be apparent. As the control signal changes the resistance of R1, the output voltage will change accordingly. So, for that matter, will the current through the circuit. Thus in the general case, an amplifier will act on both the current and voltage relationships, just as we implied when we defined an amplifier as a device which increased the **power** in a circuit.

The effectiveness of any specific active device we might use for R1 will depend upon how much its resistance changes with changes of control signal. We earlier made the acquaintance of the factors called **alpha** and **beta**, which measure this characteristic for transistors. Vacuum tubes are similarly rated for **amplification factor** and **transconductance**. Amplification factor, sometimes called "mu" because that's the Greek letter used to represent it in many engineering textbooks, applies primarily to triodes, while "transconductance" or "G_m" (its algebraic symbol) is used with pentodes. Transconductance is a contraction of "transfer conductance," and is a measure of the change in **current** in the plate circuit which results from a change in **voltage** in the grid circuit. Thus it is a direct measure of the effectiveness of the tube as an amplifier. Amplification factor measures the ratio of voltage change in the plate circuit to voltage change in the grid circuit, which is not quite so meaningful in most cases.

In both cases, only the rated quantity is supposed to change. That is, when measuring G_m for a tube, the plate voltage is not supposed to change although the current does. When measuring mu, the current is supposed to remain constant while the voltage changes. In a practical circuit, both the voltage and current change, so neither G_m nor mu is completely accurate as an indication of effectiveness. In most cases, G_m offers a more precise measurement—but that's because the G_m of most tubes varies only slightly as plate voltage changes, so that the difference between the theoretical and the practical values is small.

We have been making a major point of the fact that an amplifier boosts the power level of a signal rather than just the voltage or the current, but only a little outside reading will show you that most amplifiers are labeled as **voltage** or

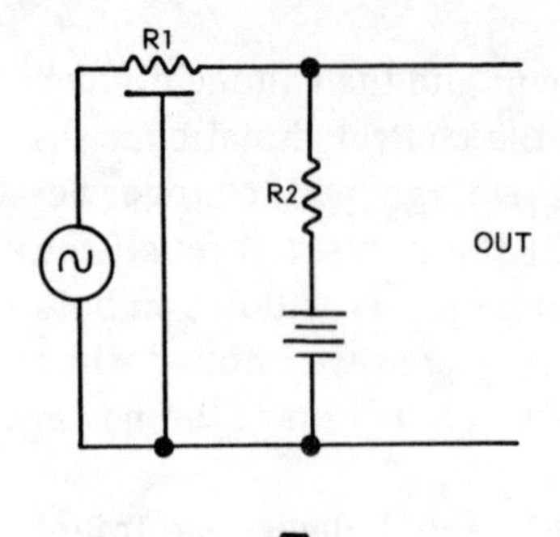

A

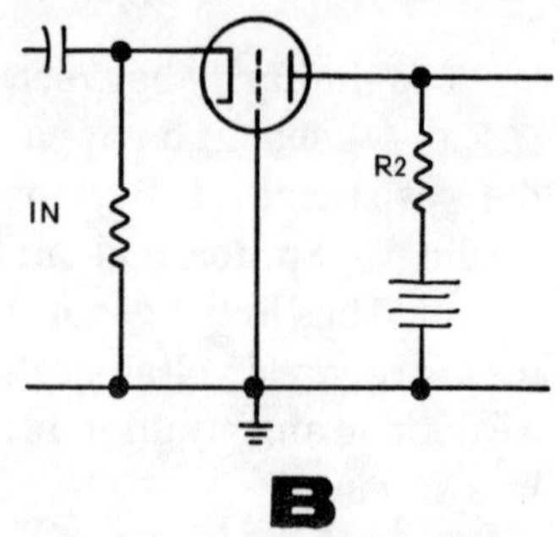

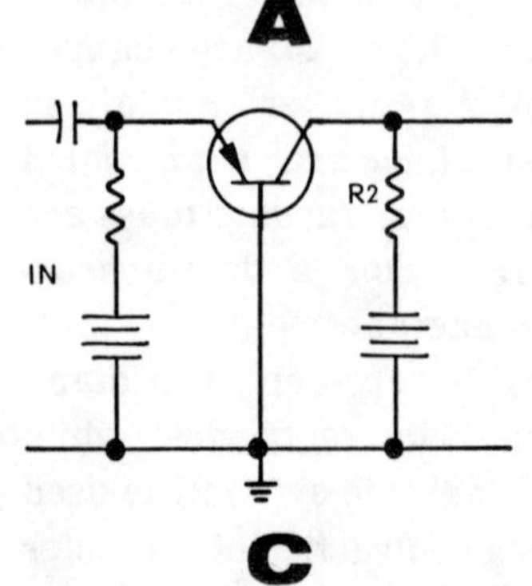

Fig. 6-2. Only voltage gain can be produced by these amplifier circuits. Output circuits are in series with input circuit, so that same current must flow through both. This precludes possibility of current gain, but difference in resistance of R1 and R2 permits high voltage gain. Vacuum-tube version (B) is normally used only at radio frequencies, to avoid need for neutralization, but transistor circuit (C) was the original circuit used for transistor amplifiers, and for several years was the only one known.

current amplifiers, with the term **power** amplifier reserved to mean one that delivers several watts of output power. How come?

We've been stressing the point that most amplifiers are actually amplifiers of power, to emphasize that they're all basically the same. The circuit of Fig. 6-1 can be used to describe action of virtually any amplifier you can find in practice, no matter how it's labeled.

It's true, though, that in the practical world, we do talk about amplifiers as if they were somehow different from current amplifiers, and both of them still separate from power amplifiers. This is simply manifestation of human nature, using verbal shorthand at the expense of absolute precision. The voltage amplifier we speak of so casually is one which is intended primarily to boost the voltage level, and any increase in power is purely a bonus. Similarly, a current amplifier is one used primarily to boost current levels, and any voltage boost is an unexpected extra. A power amplifier is one which is intended to produce appreciable power gain, with little regard for what happens to either voltage or current along the way.

In fact, some circuits cannot boost the current of the signal, and others cannot increase the voltage. Both, however, qualify as amplifiers, because they increase the power level by putting all the increase in either the voltage or current, respectively.

For instance, the circuits of Fig. 6-2 are all basically the same, and all of them serve only as voltage amplifiers. The circuit (A) shows that the input and output signals are in series with each other, and we learned in Chapter 4 that a series circuit requires all the current flow through any element to flow through all elements. Thus, every bit of current flowing in the output circuit must flow through the input circuit as well; that means that the current levels must be the same in both the input and output circuits, which rules out the possibility of current amplification.

However, if the resistance of load R2 is high compared to that of R1, the voltage drop across R2 will be high in comparison to that across R1, so that the same change in current will produce an output voltage much higher than the input voltage. This is voltage amplification.

And since power equals voltage times current, if the output voltage is larger than that at the input, and the currents are the same at both output and input, then the output power must be greater than that at the input, so we have power amplification also. Still, in practice, we call circuits like this one voltage amplifiers to indicate that they do not boost the current.

The vacuum-tube version of this circuit (B) is known as the **grounded-grid** circuit. It's most often used to amplify high-frequency radio signals, because the grid serves as a screen between input and output circuit and permits the tube to operate reliably at frequencies difficult to reach with other circuits. Some UHF triodes are specially built for use in circuits like this, with the grid brought out to a ring at the center of the tube, the plate at one end, and the cathode at the other end.

The transistor version (C) is called the **common-base** circuit; it was, historically, the first type of circuit used with transistor amplifiers. In this circuit, some of the current in the input signal can bleed off through the base connection and fail

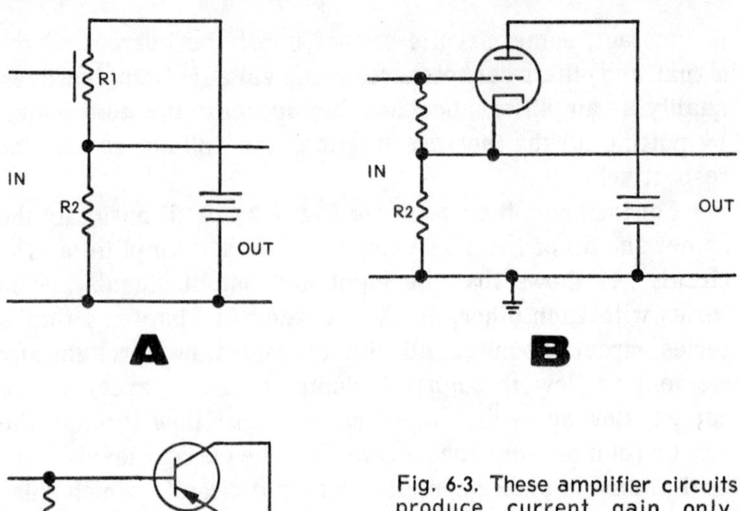

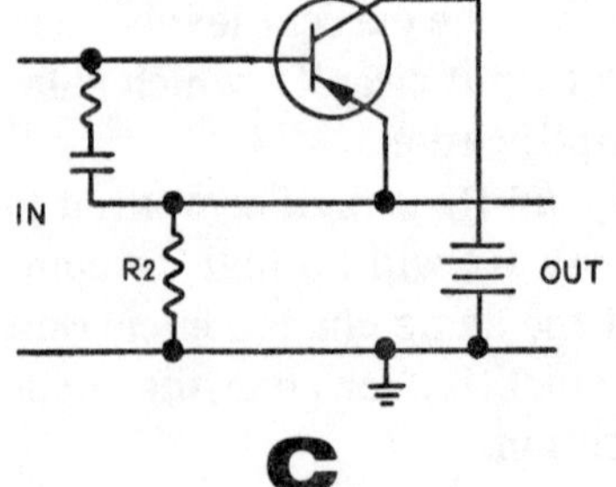

Fig. 6-3. These amplifier circuits produce current gain only. Voltage gain is not possible, because the output signal is effectively in parallel with the input signal. They are used primarily for impedance matching, because effective input impedance is much higher than actual resistance values would indicate.

to reach the output circuit. The alpha rating of a transistor measures what part of the input current goes on through to the output in this circuit; if alpha is 0.998, then 99.8 percent goes through and only 0.2 percent is bled out at the base to control action of the transistor. In this case there's an actual current loss, but the voltage gain more than makes up for it, so that the power is increased and the circuit is still an amplifier.

Similarly, the circuits shown in Fig. 6-3 are all current amplifiers. The resistor circuit, A, shows that all the output voltage appears in the input circuit where it cancels out most of the input, so that no voltage gain can result. Still, relatively small changes in input current can cause large variations of output current, so we achieve current amplification although we suffer voltage loss. As in the voltage amplifiers of Fig. 6-2, the gain more than makes up for the loss, so we have power amplification.

The vacuum tube version (B) is known as a **cathode follower**, because the output voltage taken from the tube's cathode follows the input voltage faithfully.

The transistor version (C) is called an **emitter follower** by analogy with the cathode follower, and is also known as a common-collector circuit.

The cathode and emitter follower circuits are widely used to isolate circuits from each other, because they have high input impedance and low output impedance. We'll look at them in more detail when we examine feedback a little later in this chapter.

Figure 6-4 shows the most common amplifier circuits, which produce both current and voltage gain, and can therefore be tailored by their designers to serve almost any purpose.

The resistor circuit (A) is identical with that shown in Fig. 6-1 as the generalized amplifier circuit. The vacuum tube version (B) features input going to the grid with output taken from the plate. Resistance coupling is shown, with capacitors to couple ac signals in and out. This is the normal vacuum-tube amplifier circuit, and is called a grounded-cathode circuit when it becomes necessary to distinguish it from the other two

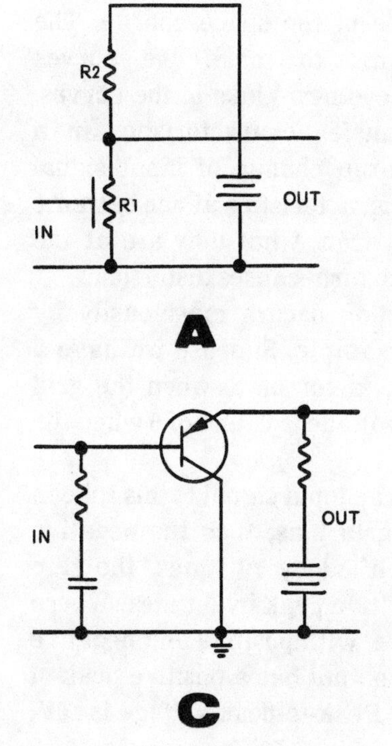

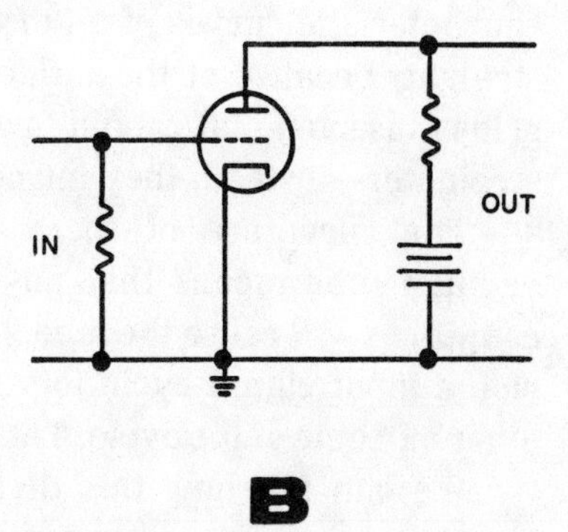

Fig. 6-4. These are the most commonly used amplifier circuits, providing both current and power gain. Both the tube version (B) and the transistor circuit (C) are basically a controlled voltage divider as shown at A. Input signal controls resistance of active device, which in turn controls divider action to valve power from steady supply. These circuits are simplified to show the principles; they omit coupling capacitors, bypasses, and biasing arrangements.

circuits. Usually, a resistor is found in the cathode circuit also, to produce proper grid bias. We'll get into that in just a bit.

The transistor version (C) is now the most common circuit used in transistor amplifiers also. Input goes to the base and output is taken from the collector. While separate batteries are shown providing bias to the two junctions, the polarity requirements of this common-emitter circuit are such that only one battery is required. A resistor can be run from the collector's power supply to the base, thus eliminating one battery. This whole business of biasing for amplifiers is the next thing we're going to examine.

We learned in Chapter 5 that transistors, to operate properly, require forward bias on the base-emitter junction and reverse biasing of the junction between collector and base. For quite different reasons, vacuum tubes also require biasing in most applications. This requirement for biasing is important enough to spend some time on it, since it is directly connected with some of the more important characteristics of amplifiers.

When we first met vacuum diodes, we found that the characteristic curves of all of them were curved rather than straight, because of the influence of the space charge. The grids inserted later did nothing to make the curves straighter—instead, they introduced new kinks in the curves.

The curvature of these transfer characteristics in a vacuum tube means that any large change of input signal conditions will cause the tube's characteristics at one extreme of the input signal cycle to vary from what they are at the other extreme of the cycle. This, in turn, causes distortion.

We can see how this distortion occurs most easily by plugging some figures into an example. Suppose we have a tube which has an amplification factor of 10 when the grid voltage is −1, of 9 when the grid voltage is 0, and of 6 when the grid voltage is +1.

If we apply a 2V peak-to-peak ac input signal to this tube in a conventional circuit, with no grid bias, then the negative peak of this signal will be amplified by 10 times, the zero crossings by 9 times, and the positive peak by 6 times. Where the input was a smooth sine wave, with positive and negative peaks each 1V from ground, the output has a positive peak of 6V and a negative peak of −10V. Peak-to-peak voltage is 16V, so the positive and negative peaks can no longer be symmetrical (if they were, zero crossings would occur halfway between, so that the positive peak would be 2V greater and the negative peak 2V less negative). The fact that the output signal is not a replica of the input is what we call **distortion**. Any time the characteristics of an amplifier change from one point in the input signal to another, distortion must result, and in most amplifier applications distortion is something to be avoided like the plague.

Our sample tube, to continue the example, might have an amplification factor of 10 when the grid voltage is anywhere between –1 and –3 volts. In this case, putting a –2V negative bias on the grid would eliminate the change in characteristics, because our 2V peak-to-peak input signal would now swing the actual grid voltage from –1V (the –2V bias, plus the +1V positive peak) to –3V (–2V bias plus –1V negative peak), and the output would then be a 20V peak-to-peak signal in which both halves of the cycle were symmetrical. Distortion would disappear.

The purpose of bias voltage on a vacuum tube, therefore, is to establish an operating point which permits the circuit to have the desired characteristics. In our example, the bias set an operating point which kept the amplification constant over the 2V input signal's range of amplitude, and so eliminated distortion. In other circuits, different operating points might be desired. Strange as it sounds, in some circuits we might even deliberately choose a bias voltage that would result in extreme distortion. That's what we do to achieve high efficiency in many radio-frequency power amplifiers, because power conservation is more important than the prevention of distortion in that precise application.

In a transistor circuit, some bias voltage is required in order to establish amplifier operation, but the exact value desired in any specific circuit is chosen for the same reason and in much the same manner as is the bias voltage for a tube. Choice of an operating point is one of the biggest decisions a circuit designer has to make, and maintaining the desired point in operation is one of the major tasks faced by maintenance personnel in many cases.

Fortunately, design tricks exist which make it possible to set up a circuit which adjusts its own operating point, and most of the "plug in and play" electronic equipment sold to the public uses such circuits. That's what keeps you from having to adjust bias on every amplifier stage every time you turn on your radio or TV set.

We'll get back to these **automatic bias** circuits in a little while. They are based on the principles of feedback, though, so we will have to find out how that works first. And before we delve into feedback, we have a few things to finish.

A half-dozen or so paragraphs back, we mentioned in passing that in most amplifier applications distortion is something to be avoided like the plague. That's not, strictly speaking, a totally true statement. Many amplifier circuits deliberately introduce distortion for one reason or another. Most of them, however, strive to avoid distorting the output signal as finally delivered.

For instance, a radio-frequency (rf) power amplifier uses a mode of operation which produces extreme distortion of the individual rf cycles, in order to achieve greater power efficiency. However, the rf cycles which are output from the circuit are essentially undistorted, because the tuned circuit which provides the amplifier's load impedance filters out all the distortion.

Another example may be found in audio power amplifiers, which frequently use two symmetrical amplifier circuits as a single output stage. This **push-pull** configuration, so called because one circuit "pushes" while the other "pulls," can tolerate severe distortion on the part of the individual circuits, because the other circuit cancels it out.

In both these examples, distortion within the circuit was severe, but the output signal was essentially undistorted. In other amplifier applications, no distortion is permitted in the circuit. These extremes represent different ways of using amplifiers, which in turn require some means of identification.

We have a number of different ways available to identify amplifiers. We've been seeing some of them in these words. For instance, we speak of audio amplifiers, rf amplifiers, and so forth, classifying the amplifier according to the frequency of the signal it is intended to handle.

This is a useful classification method, and for some purposes is the only meaningful one, because requirements for operating in a dc amplifier which must respond accurately to signals of zero frequency are most different from requirements for proper operation in the UHF range.

It does not, however, tell us much about the distinction between those circuits which distort and those which do not.

Another way of categorizing amplifiers is by the type of active devices used, such as vacuum-tube , transistor, or even hybrid (which uses both tubes and transistors). Again, for some purposes this is necessary, but it tells little or nothing about the distortion characteristics.

We could classify according to the amount of distortion introduced, with low-distortion, moderate-distortion, and high-distortion amplifiers. For these labels to have any meaning, we would have to specify how much, and what kind, of distortion it took to make the step from low to moderate, but the idea at least is a step in the right direction. The problem here is in finding a way to actually make the classification.

We really do use a rather similar classification in the area of rf power amplifiers, which are classified into the two categories of **linear** and **nonlinear**. A linear amplifier is one which does not introduce objectionable amounts of distortion into a modulated signal; a nonlinear amplifier is one which fails to meet the requirements of a linear. The dividing line between linear and nonlinear frequently is the ear of the listener; it is, however, possible to measure total distortion in the signal, and specify a percentage which is not to be exceeded.

To describe the operating-point conditions of amplifiers in general, however, three classes of operation were established many years ago. The classes often lead to confusion, because they are defined differently in different texts, but for our purposes we can consider a **class A** amplifier to be one in which the resistance of the active device in the amplifier never becomes infinite within an input signal cycle. That is, the input signal never cuts the amplifier off, nor is the device cut off in the absence of signal.

A **class B** amplifier is one in which the resistance is infinite (device cut off) in the absence of signal, but bias is overcome by the signal so that in the presence of an input signal, current flows for at least half the signal cycle.

A **class C** amplifier is one which is biased deeper into cutoff, so that current flows for less than half a signal cycle.

From these definitions and our previous discussion of distortion, it's apparent that both **class B** and **class C** operation must introduce distortion. **Class A** operation, also, may introduce distortion—our original example showing the need for bias voltage was operating class A, but distorted the signal severely. Thus, class A operation is necessary, but not necessarily sufficient, to avoid distortion.

However, class B operation can be used with push-pull circuits in which one device supplies the load while the first is cut off, and vice versa, to provide low-distortion amplification. Most hi-fi amplifiers use some variation of push-pull amplification. With transistors, it may show up as a "totem pole" output, but it's still a sharing of the effort between two or four symmetrical circuits.

Note that the one essential variable which determines whether any specific amplifier is class A, B, or C is its bias. This means that you can change from class A to class C operation merely by changing the bias. In most cases, other circuit changes would also be necessary to optimize things, but the class of operation is established exclusively by the biasing.

This makes it possible, with some automatic biasing circuits, to make a circuit adjust itself from class A to class B. Some push-pull circuits do this; they are biased so that small signals are handled in class A. Larger values of input signal automatically increase the bias, switching to class B operation. These circuits are sometimes called class AB, to show that they are a mix between classes A and B. The AB is also used for other purposes, though, so don't trust it every time.

FEEDBACK

In the early days of electronics, back when it was still known to most folk as "radio," **feedback** was a horrible condition—to be avoided like bad breath or body odor.

Actually, though, feedback in its most general sense is probably one of the most basic ideas in existence. One example is the idea of "cause and effect"; any time the effect feeds back to modify the cause, we're seeing feedback at work. When you reach out to turn a page in this book, feedback makes it possible. Your eyes tell you that the page is ended, you move your hand, and your eyes measure the distance from hand to book to tell you which way to move the hand and when to open the fingers. This feedback of information from eye to hand results in the page getting turned, to provide fresh input to the eye.

What the oldtimers were really objecting to under the name of "feedback" was **uncontrolled** feedback. Just like anything else, feedback running out of control can only produce unpredictable results. With proper control, though, feedback is essential to modern electronics.

For instance, when the microphones and speakers of a PA system are badly placed, so that the mikes pick up output from the speakers and produce a loud howl, that's uncontrolled feedback running wild.

Similarly, an unneutralized triode amplifier will oscillate with extreme enthusiasm when its input and output circuits both contain resonant "tanks" tuned to slightly different frequencies, due to uncontrolled feedback through the tube's plate-to-grid capacitance.

But the neutralization which makes it possible to use the tube anyway is actually only more feedback, tightly controlled so that it cancels out the bad effects of the original unwanted feedback.

And the reason that hi-fi amplifiers are able to achieve unbelievably small distortion ratings is because their designers took advantage of controlled feedback to cancel out distortion introduced by the amplifier circuits.

Figure 6-5 shows the basic principle of feedback in its most simplified form. A little of the output of an amplifier is tapped off and fed back to provide an additional input signal. If the feedback signal is opposite in phase to the original input, we have negative or **degenerative** feedback; if it is in phase with the input, the feedback is positive or **regenerative**. Both kinds have their uses, and neither kind is desirable unless it is controlled.

An ac signal of steady frequency can be completely characterized by specifying its amplitude and its phase. As we

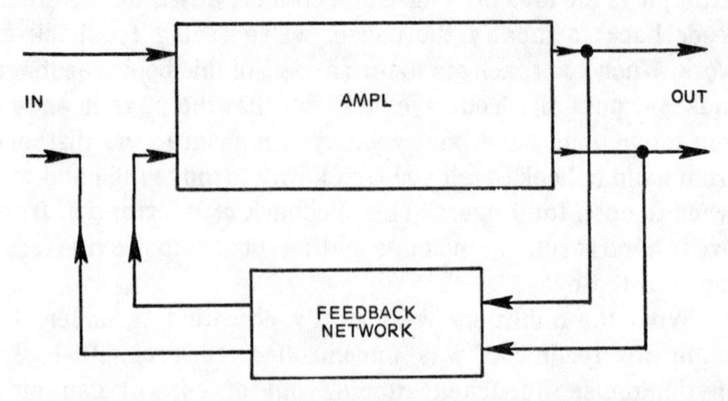

Fig. 6-5. Basic principle of feedback is that a part of the amplifier's output signal is either added to, or subtracted from, the input signal. If added, it's positive feedback; if subtracted, negative. Any distortion in the output which was due to the amplifier will not be in the input, so that negative feedback selectively reduces gain for distortion products (by introducing **counter distortion** into input signal). Feedback also permits "servo" action, in which circuit adjusts itself to meet changes in operating conditions.

examine the principles of feedback, it's easiest if we work only with single-frequency sine-wave ac signals of constant amplitude (strength), since any changes we find then must be due to the effects of the feedback. We can then extend our findings to more usual mixed-frequency signals with varying amplitude.

When two different ac signals combine (assuming that both are sine waves of the same frequency) they produce a single signal of the same frequency, but the phase and amplitude of the composite signal depend upon the relative phases and amplitudes of the original pair of signals.

If signal A is of 5V peak amplitude, and signal B has 10V peak amplitude but is 180 degrees out of phase with A, the final signal will be 5V peak amplitude and will have the phase of signal B. In effect, signal A has canceled out half of signal B's amplitude, and disappears in doing so, with the result that only half of signal B's strength survives to become the composite.

With the same peak amplitudes but with both signals in phase, the composite would have 15V peak amplitude and original phase. At every instant throughout the cycle, the two would add together to form a composite equal to the sum of the original signals.

If the phase difference is anything other than 0 or 180 degrees, the composite signal's phase will be different from either of the original signals. If two signals of identical amplitude but 90 degrees apart in phase are combined, the result is a signal which has 141.4 percent of the original amplitude, and is midway between the original signals in phase. This happens because as each original signal crosses zero, the other signal is at one of its peaks; at that instant, the composite signal's value is equal to the peak amplitude of the input. A little later (45 degrees of phase, in fact) the two input signals are at identical absolute voltage values. Depending upon just where in its cycle each signal is, the input amplitudes may both be positive, both negative, or may be of opposite polarity.

When they are of opposite polarity, the inputs cancel each other and the composite output signal crosses zero. When they are of the same polarity, one is rising and the other is falling, so that the composite reaches its peak value for that polarity. Since the instantaneous amplitude of a sine wave 45 degrees either side of its peak is 70.7 percent of the peak value, and the composite peak is the sum of two such instantaneous values, the composite's peak is 141.4 percent of either input signal amplitude.

It's important to know how ac signals combine, but for the purpose of understanding feedback action the only essential phases are 0 degrees and 180 degrees. Feedback signals having zero-degree phase relation to the original input signal provide positive feedback, and those with 180-degree phase provide negative feedback. When we provide controlled feedback with the circuit shown in Fig. 6-5, we control both the amplitude and the phase of the feedback signal, and so control the entire feedback situation. Note that accidental feedback—the kind the oldtimers detested—cannot be controlled.

Before we look at the reasons for using feedback, let's see what its effects are when we put it into an amplifier.

Let's assume that our amplifier has a gain of 10, so that it produces a 10V output signal with a 1V input, and let's arrange to feed back 10 percent of the output signal to the input 180 degrees out of phase, so that the feedback is negative.

Before the feedback is connected, a 1V input signal would produce 10V of output. If we feed back one tenth of this out of phase, we are feeding back a 1V signal—which would exactly cancel the input signal, leaving nothing to be amplified. Yet with no input, there is no output, so there would be no feedback to cancel out the signal. Attempting to analyze feedback by this approach invariably leads to a circle of contradictory confusion such as we have just entered.

The apparent contradiction doesn't actually happen in the real circuit, because a finite amount of time (even though it's so few billionths of a second that we can usually ignore its existence) is required for a signal to change value or to get around the feedback loop.

For example, when we apply the 1V signal to the amplifier's input after connecting the feedback circuit, the actual voltage level at the amplifier input cannot become 1V instantly. Since it was previously at zero, it must go through all the values between 0 and 1V before reaching 1V.

With most signals, this occurs in millionths of a second and we usually ignore it. In the case of feedback, signals get around the feedback loop from input to output and back to input in billionths of a second, and a change which takes a millionth of a second is a thousand times longer than the circuit takes to act. We can no longer ignore the rise time of the input signal.

As the signal rises from zero, it must pass through the 0.1V (100 mV) level. As it does so, the output is 10 times greater (1V). We feed back 100 mV of this to the input, but by this time the input signal has risen some more, so is not entirely canceled. If the input is up to 150 mV by this time, only 100 mV is canceled, leaving 50 mV to be amplified.

This 50 mV produces an output 10 times greater, or 500 mV, and we feed back 10 percent of it, or 50 mV. By this time, though, the input signal has risen still higher than the 150 mV which established the 500 mV output level.

When output level gets up to 5V by this process, the feedback voltage has risen to 500 mV, or 0.5V. If at this time the input signal has reached its final amplitude of 1V, the 0.5V of feedback from a 5V output level will leave 0.5V of input signal, which, when boosted by 10 times, produces the same 5V output. Thus, the output level never rises higher than 5V for a 1V input signal.

The net effect of our feedback arrangement was to reduce the amplifier's gain from 10 to 5.

Negative feedback always reduces the gain of an amplifier. In so doing, it also reduces the effect of any distortion which the amplifier introduces into the signal, because none of the distortion is canceled by any part of the input signal. Thus, the feedback has more effect upon distortion than it does on the signal, which reduces the percentage of distortion in the output. This is one of the main reasons for using it.

By using enough negative feedback, gain can be reduced almost to zero, but can never be completely cut off. The cathode- or emitter-follower circuits are examples of this; they introduce 100 percent negative feedback of output voltage, and as a result their voltage gain is always less than unity (which means they actually introduce a voltage loss, although it may be very small). Similarly, the grounded-grid and common-base circuits have 100 percent negative feedback of output current, and so their current gain is always less than unity.

Let's take our previous example, and change the amount of feedback in it. Rather than feeding back 10 percent of the output voltage, let's use only 1 percent—but we'll feed it back with a 0-degree, rather than 180-degree, phase difference, to make it positive or **regenerative** feedback.

Now when we put in a 1V input signal, the output is initially 10V. The feedback takes 1 percent of this, or 0.1V (100 mV) back to the input where it increases the input signal to 1.1 volts, which in turn increases the output to 11V. This brings the feedback amplitude from 0.1V up to 0.11V, raising the input to 1.11V and the output to 11.1V. Every increase in the input brings the output up by 10 times the amount of the increase, and every tenfold increase of output adds 0.01 more to the input.

It would appear that this process could never end, and that the output voltage would keep rising. In principle, that's true, but in practice it ends when the increase in output level is less than the normal "noise" fluctuation of several millionths of a volt, which is always present in any circuit.

Even in theory, though, the sum of the infinite series the loop produces is limited to 1.11111111111..., because the amount of input increase is smaller than the product of the feedback fraction multiplied by the gain. That is, every increase of output produces less effect on the output than did the one before.

Thus the final level is 100 / 9, no matter how long we let the loop go.

If we increase the feedback fraction to 0.05, we get similar results, but with much greater increase in amplifier gain. A 1V input gives 10V out, which increases the input by 0.5V to 1.5V and produces 15V out. This, in turn, brings feedback up to 0.75V and raises the input level to 1.75V, producing 17.5V output. This gives a feedback voltage of 0.875V, and effective input of 1.875V for an output level of 18.75V.

No matter how far we follow this loop, though, the output will never pass 20V. It won't even reach 20V in any finite number of trips around the loop, but it will get so close in a few thousand passes that you won't be able to tell the difference. And since the loop delay time is only a few billionths of a second, that "few thousand passes" will take considerably less than one ten-thousandth of a second to accomplish.

We can prove rather readily that the 20V output cannot be exceeded, by assuming that we have reached it. Since the feedback fraction is 0.05, our feedback voltage for a 20V output would be exactly 1V. This 1V of feedback, added to the 1V of original input, would produce 20V output—but that's what we started with, and there's no increase. Therefore, the output could not rise above this level.

If we increase the feedback fraction to 0.099 (and take extreme care to assure that it doesn't inch on up to 0.1), we get gain that's difficult to believe. Without going through the loops to prove it, we'll just say that effective gain in this case is 1000, or 100 times that of the amplifier without feedback.

If we bring the feedback fraction up to 0.0999 the gain is increased by 1000 times to 10,000. With a positive feedback fraction of 0.09999, the gain would be 100,000, and so forth.

But if we ever let the gain get up to 0.1, things change. Now the limit on increase of output vanishes. A 1V input would produce 10V output, which would give another volt of input. This would produce 20V output, giving 2V feedback and raising input to 3V. The process would, in theory, go on forever without any limit. In practice, supply voltage availability puts a limit on output. When the circuit is producing all the output the power source can supply, it cannot increase any more.

Something else happens, too. Let's assume that we put just one half-cycle of 1V input into such a circuit, and then took the original input away. That half-cycle would produce a similar half-cycle at the input, of the same strength as the original one. If we gave it one full cycle to begin with, and the frequency of the signal was such that the feedback signal made it around the loop back to the input just as we took the input away, the feedback signal would take the place of the original input and our amplifier would now supply its own input signal. It would continue to produce 10V of output, without increase, because that would be just enough output to provide a 1V input signal through the feedback path.

Such action is called **oscillation**, and that's how an oscillator works to turn dc into ac at almost any frequency you can imagine. An oscillator circuit consists of three major parts; one, the amplifier, provides gain. The second, the resonator, determines the frequency of oscillation. The third, the feedback network, keeps oscillation going by providing input to the amplifier. Any time the feedback is positive and the product obtained by multiplying amplifier gain times feedback fraction is one or more, the circuit will oscillate.

A circuit having positive feedback but with the feedback fraction kept small enough so that it produces a product less than one when multiplied by amplifier gain is said to be **regenerative**. In the early days of radio, the regenerative receiver boosted amplifier gain to fantastic figures and let one expensive vacuum tube do the work of several. These receivers always featured a regeneration control which required tricky adjustment to get the desired gain while avoiding oscillation. What the control did, of course, was to vary the feedback fraction so that it could keep "loop gain" in the feedback path less than one, producing regeneration without oscillation.

By now, you're probably wondering how to figure out the effective gain of a feedback amplifier without going through all the loop figuring we've done in our examples. Fortunately, there's a simple formula for determining it, which works with both positive and negative feedback. To use the formula, you need only the gain of the amplifier without feedback, and the **feedback factor** of the amplifier. The sign of the feedback factor determines whether it's positive or negative feedback; for degenerative feedback, the feedback factor is negative, and for regenerative feedback, the factor is positive.

To use the formula, first multiply feedback factor by amplifier gain to obtain a **loop-gain** figure. If this number is positive and equal to (or greater than) 1.000, the circuit will oscillate. Otherwise, subtract the loop gain figure from 1.000, and divide the original gain figure by the remainder. The result is the gain of the amplifier with feedback.

For instance, our first example was an amplifier with a gain of 10 and negative feedback of 0.1. This gives us a loop gain of 10 times –0.1, or –1. Subtracting –1 from 1 gives us 2, because subtracting a negative number results in addition, and the original gain of 10 divided by 2 gives us a final gain of 5.

If a cathode-follower circuit would have gain of 200 before feedback, the 100-percent negative feedback introduced by the circuit would give a loop gain of 200 times –1, or –200. Subtracting –200 from 1 gives us 201, so final gain would be 200/201 or just less than 1.0.

When the feedback is positive as in our first example of regeneration, the gain was 10 and the feedback was 0.01. Loop gain was 10 times 0.01, or 0.1. The subtraction gives us 1 — 0.1, or 0.9 and dividing the original gain of 10 by 0.9 gives us a final gain of 100/9.

Bringing the feedback factor up to 999/10000 gives us a loop gain of 0.999, which leaves 0.001 when subtracted from 1. The original gain of 10, divided by 1/1000, is increased a thousand times to 10,000.

If we have a circuit which contains both positive and negative feedback, they will tend to cancel each other. Only that remaining part which is not canceled can affect the circuit's operation. Thus, if we have accidental positive feedback which is causing an amplifier to oscillate, we can put in enough negative feedback to reduce the loop gain to something less than 1.0 and prevent the oscillation. If we exactly balance the accidental feedback, we have neutralized the circuit.

We mentioned "automatic bias" earlier, but pointed out that an understanding of feedback was necessary to comprehend how such circuits work. Now that we've seen how feedback can control the gain of amplifiers, we're ready to see how it can control other factors such as bias.

There is a big difference between the feedback we've just been studying and such circuits as those used for automatic bias control. The feedback we've looked at so far is put in to keep operating conditions constant by stabilizing gain and reducing distortion, while the "servo" circuits we're going to examine are used to automatically modify operating conditions and thus account for unexpected changes. In practice, the biggest actual difference is that the ac signals involved in the feedback loops we've seen so far are replaced by slowly varying dc, sometimes called an "error" signal.

One of the simplest of these servo circuits is that used to provide automatic grid bias on a triode vacuum tube. The circuit is shown in Fig. 6-6, and this is how it works:

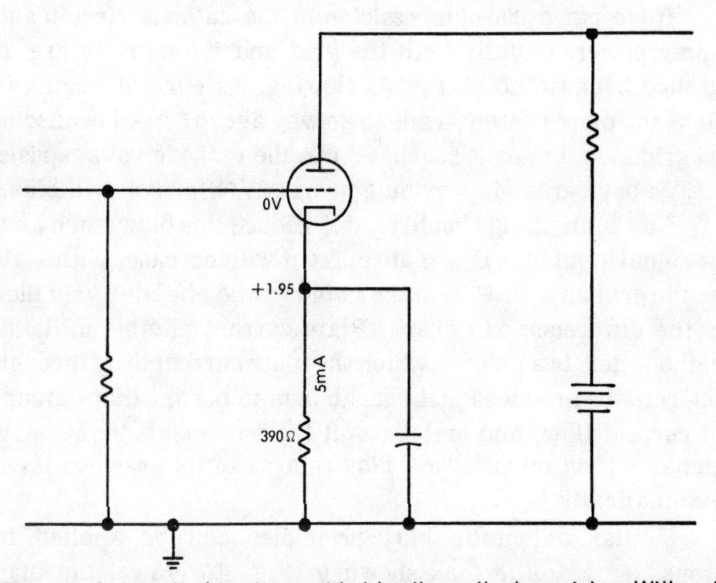

Fig. 6-6. Automatic bias is provided by the cathode resistor. With grid grounded for dc, bias between grid and cathode is set by cathode voltage. Cathode voltage, in turn, is established by drop across resistor, so that as current goes up, the bias voltage increases, which tends to cut current back. This servo action rapidly drives current flow to the level which provides the designer's intended value of grid bias, and holds it there even though supply voltage may change, or components change characteristics with aging.

The cathode resistor provides grid bias, because the effective bias voltage is that potential existing between grid and cathode, and it doesn't really make any difference which of them is connected to ground. By connecting the grid resistor to ground, we clamp the grid voltage at zero while permitting ac to vary it around zero for the incoming signal. The cathode is positive to ground by the amount of the voltage drop across the cathode resistor, so (by looking at it backward, so to speak) the grid must be negative to the cathode by the same amount. The capacitor in parallel with the resistor keeps the cathode grounded so far as the ac signal is concerned. This trick of keeping ac and dc grounds separate, incidentally, is in wide use.

Let's assume that we have a tube which requires a —2V grid bias to have 5 mA of plate current. Ohm's law tells us that we must have a 400-ohm resistor to produce a 2V drop with 5 mA flowing, and the nearest standard value of 390 ohms will give us 1.95V at 5 mA.

If we put a 390-ohm resistor in the cathode circuit and apply power, initially both the grid and the cathode are at ground because no current is flowing. As current begins to flow, the plate current tends to go way above 5 mA because of no grid bias, but as it reaches 5 mA the cathode voltage rises 1.95V above ground, providing nearly 2V effective grid bias.

Now if anything should tend to reduce this bias (such as a dc signal input of +1V), plate current will increase. With +1V on the grid and +1.95V on the cathode, the effective grid bias is the difference or —0.95V. Plate current climbs until the cathode gets to a point at which the plate current drop through the resistor produces just enough bias to permit that amount of current flow, and stabilizes at the new point. If the +1V signal is then taken away, bias returns to its previous level automatically.

Similar automatic bias principles can be applied to transistor amplifiers as shown in Fig. 6-7. Here, the bias control resistor is the one in the emitter circuit. The voltage divider which feeds the base resistor establishes a voltage level for the base, and the voltage drop through the emitter resistor then holds the emitter just far enough from that voltage level to provide the right bias to maintain emitter current steady. By using this principle, designers can avoid most of the critical adjustments of bias which were once required.

In the space available here, we've only been able to scratch the surface of the principles and possibilities which feedback involves. Hopefully, though, this will keep you from becoming too confused by it as you progress through practice, and will help you determine what's going wrong in the meantime.

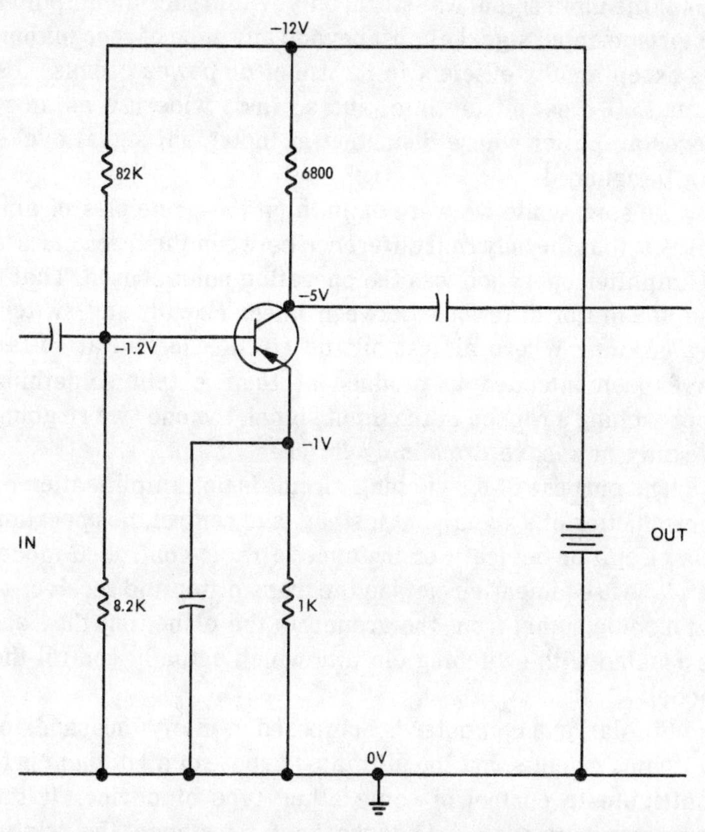

Fig. 6-7. Automatic bias circuit for transistor amplifier makes use of voltage divider shown here as 8.2K and 82K resistors in series, in base circuit, to fix base voltage at 1.2V negative from ground. With 0.2V drop across base-emitter junction (which is a typical value for a germanium transistor), this sets 1V as the desired emitter voltage. The 1000-ohm emitter resistor requires 1.0 mA current through it to develop this voltage, so only that much current can flow. Collector voltage is established at approximately 5V by 6.8V drop through collector load resistor.

SWITCHING CIRCUITS

We've already made the acquaintance of a few switching circuits, although they haven't been called that. The first was the latching relay, in Fig. 5-2. This one switched an external circuit on, then held it until receiving another signal telling it to break the circuit.

Another switching circuit was the **class C** amplifier which we skimmed over lightly in the first part of this chapter. The class C amplifier is one which is normally biased deeply into cutoff, and is switched into conduction by the positive-going tips of the input signal waveform only. While such an amplifier distorts its input signal almost beyond any hope of recognition, it's exceptionally efficient in its use of dc power because it's turned off most of the time, and so finds wide use as an rf power amplifier where distortion of individual signal cycles can be ignored.

We saw, while we were examining the principles of amplifiers, that the only real difference between the three classes of amplifier operation was the operating point chosen. That's also the major difference between **linear circuits** and **switching circuits**. Where almost all the circuits looked at so far have been intended to produce at their output something approaching a replica of the input signal, the ones we're going to study now have drastically different intent.

The purpose of a switching circuit is not amplification or reproduction of a signal, but instead is to control the operation of a circuit or device. For instance, a radio-controlled model airplane uses linear circuits in the transmitter and receiver to get a radio signal from the ground to the plane, but they are associated with switching circuits which actually control the model.

Similarly, a computer is composed of many thousands of switching circuits, but the ultimate purpose of all of them is to contribute to control of some other type of device. If the computer is printing our paychecks, for instance, the printer is the device being controlled. If it's monitoring the launch of a space probe, the rocket motor may be the device being controlled.

But regardless of the circumstances, the purpose of the switching circuit is always to control the operation of something else, and to do so it need use only the two states known as **off** and **on**.

When we speak of a switch being off or on, no explanation is necessary. If it's off, current cannot flow through it, and if it's on, the reverse is true.

The off and on states of a switching circuit are similar. Most of today's circuits use semiconductors, because they require less power and any practical control circuits require vast numbers of devices. The on state is that which we have already met as **saturation**, and the off state is the other extreme of **cutoff** (or pinchoff, if we're talking about FETs).

Since a transistor amplifies power, even when it's used as a switch, we can do all our controlling at low power levels, then control bigger and bigger output transistors to turn on and off the final controlled device. For instance, the computer which is printing paychecks ends up controlling rather powerful electromagnets which drive a hammer against the check to print the desired figures, but most of the switching occurs at power levels in the milliwatt region (today's circuits usually use 5V power supplies, with possibly 10 mA being drawn by each switching element) even though the hammer magnets require several amperes to drive them properly.

Once the desired switching has taken place at low power levels, a voltage is placed on an input line which in turn feeds a **driver**, turning it on. The driver is a power transistor capable of controlling the necessary current, and of being switched on or off by the output of the low-level switches.

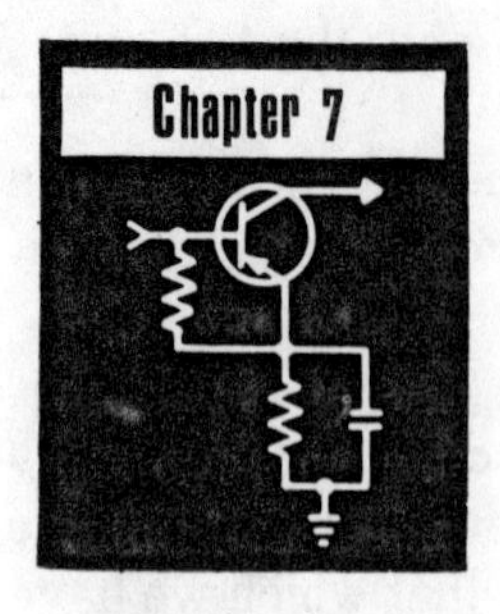

Learn By Doing

In the preceding six chapters, we've been examining and discussing the elements of practical electronics—but the activity has all been mental.

The best way of learning anything, though, is by doing something with it. That's why schoolteachers assign homework; doing the assignments provides the necessary practice to reinforce and complete the learning process.

Presumably, you're interested in electronics already and have no need for assignments and homework; however, the simple projects described in this chapter have been selected for their ability to reinforce the principles set forth in the preceding chapters so that they will serve the same purpose. You can make your own assignments, as your interest dictates, and have fun building things while learning how and why they work.

Incidentally, don't be surprised if your first projects fail to operate as you expect them to. Practicioners of the art of electronics learn early that **Murphy's law**—"If anything can possibly go wrong, it will!"—applies 'round the clock, and there are always things that can go wrong. Take heart from the fact that we learn more from our mistakes than from our successes, if we take the time and effort to determine just what did go wrong and why, and you'll soon join the ranks of the electronics experts.

A MINIPOWER OSCILLATOR

Transistors have just about driven the vacuum tube out of electronics, except for a few specialized jobs which the tube still keeps a grip on (such as providing the TV picture). This project, the first of our learn-by-doing set, shows several of the reasons.

The only parts required are a transistor, a transformer, a dime, and some wire. Later we'll replace the dime with an orange or a lemon and a couple of nails!

The project itself is an audio oscillator, consisting of the transistor and the transformer. The dime, or the fruit, provides the power source, which serves two purposes. First, it models the early voltaic pile (Chapter 2) which powered most of the basic discoveries of electronics; second, it demonstrates how very little power the transistor requires for operation.

For the transistor, one of the early pnp experimenter units such as the CK722 or 2N107 is recommended, although any general-purpose transistor will do. If you happen to have an npn unit, use it, but reverse battery polarity from that described here.

To save wear and tear on the transistor leads, punch three holes through a piece of thin cardboard (such as a matchbook cover or posterboard) with a pin or needle, and thread the leads through them. You can then write the identity of each lead near its hole, keeping them straight. Figure 7-1 shows which lead is which for most popular transistor case styles.

With the transistor safely mounted on cardboard, connect it to the transformer as shown in Fig. 7-2. Any "universal output" transformer can be used. Also connect the headphone to the transformer at this time.

Now for the power source. Put a piece of blotting paper or felt against the dime, and moisten it with saltwater. Hold one wire against the coin, and press another wire against the

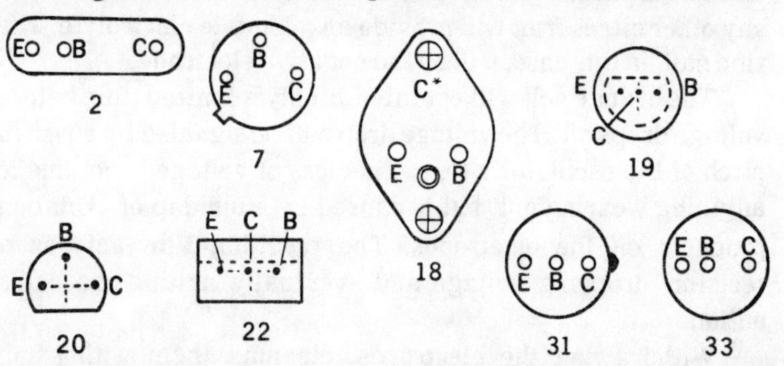

Fig. 7-1. Base connections for most popular types of transistors used by home experimenters are shown here. These show leads as viewed from the bottom of the transistor.

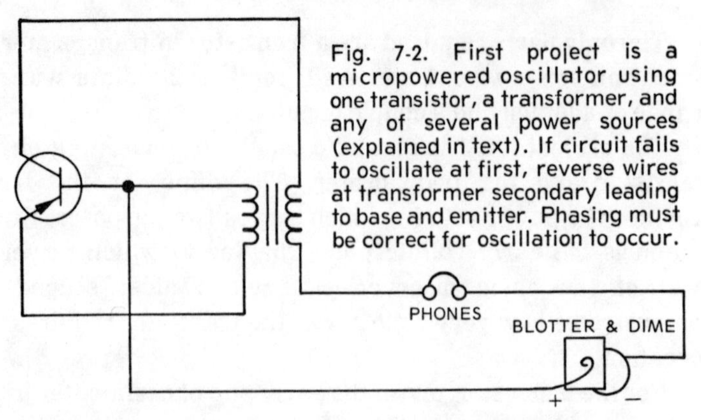

Fig. 7-2. First project is a micropowered oscillator using one transistor, a transformer, and any of several power sources (explained in text). If circuit fails to oscillate at first, reverse wires at transformer secondary leading to base and emitter. Phasing must be correct for oscillation to occur.

electrolyte-moistened paper. The wire touching the paper will be the negative lead.

If no tone is produced in the headphone when the "battery" is connected to the circuit, reverse the connections of the transistor base and emitter to the transformer and try again. The feedback phase must be positive for the circuit to oscillate, and this phase is determined by these connections.

The dime-and-saltwater cell produces a potential of about 700 mV (0.7V) if a silver dime and copper wire is used; under the load imposed by the oscillator this drops to about one-half volt. A vacuum-tube circuit would require many times this much power just to light the filament, to say nothing of providing plate power, yet the transistor operates nicely.

Many other power sources can be used in place of the dime cell. For instance, if you can locate a length of copper wire or a wire nail and one zinc-coated roofing nail, these can be used as electrodes. Driving them into an orange, a lemon, a lime, or any other citrus fruit will provide an adequate electrolyte. The zinc nail, in this case, will be the negative electrode.

The orange cell will operate for only a limited time before voltage drops off. The voltage drop will be signaled by a rise in pitch of the oscillator's tone. This loss of voltage is not due to anything wearing out, but is caused by a buildup of oxidation products on the electrodes. The resulting film acts as a resistor, dropping voltage and eventually halting the cell's action.

Withdrawing the electrodes, cleaning them with sandpaper, and reinserting them will restore the cell to like-new condition. This can be repeated until the fruit dries out.

A selenium or silicon solar cell or "sun battery" will power the oscillator adequately. Normal room lighting provides good output. At night, an automobile headlight 30 feet away is ample light to produce circuit action. Since the tone changes with supply voltage (high at low voltage, and falling in pitch as voltage rises), and solar cell output voltage rises with light level up to its limiting point, this demonstration circuit can serve a practical purpose as an audible light meter.

In addition to demonstrating the action of chemical and light-driven power sources, this project shows how feedback provides oscillation. As shown in Fig. 7-2, the transistor is connected as a common-base amplifier with all of its output signal fed back through the transformer to the emitter for input. If transformer phase is wrong, nothing happens, but if the phasing is correct to produce positive feedback, oscillation is certain.

Moving the positive terminal of the power source from the base to the emitter changes the amplifier portion of the circuit from a common-base arrangement to common-emitter. This change of the return point also effectively reverses phase of the input signal, so that everything is still right for oscillation. The slightly higher gain of the common-emitter circuit, as compared to the common-base arrangement, causes the pitch to decrease slightly when the positive connection is moved to the emitter, but action is otherwise unchanged.

It's interesting to see where the base bias current comes from in this circuit. You'll remember that in Chapter 5 we discovered that the base-emitter junction of a transistor must be forward-biased, and the base-collector junction reverse-biased, for transistor action to occur. Yet this circuit applies no dc to the base-emitter junction; the power source connects to the collector circuit only.

As a result, this circuit must be operating in class C, with a bias arrangement that permits no current flow in the absence of input signal.

When the power source is connected to the circuit, the first attempt of current to flow through the transistor induces a signal in the transformer, providing the necessary input signal. This input signal in turn provides the necessary bias to

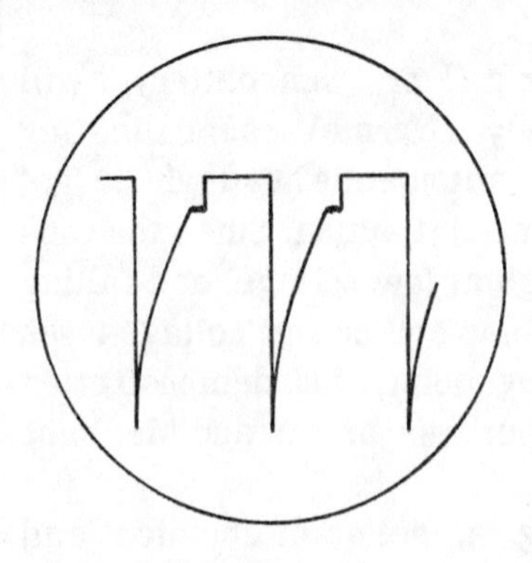

Fig. 7-3. Waveform of oscillator output. Rising portion is produced as current builds up to steady state level. When this level is reached, output jumps sharply and holds there momentarily, then is cut off abruptly as magnetic field of transformer collapses. This starts the cycle all over again, and process continues so long as current is applied.

permit continued current flow. So long as the current flow is changing, the transformer keeps operating.

When the current flow reaches its steady-state condition, the change stops and the transformer no longer produces the required input signal. The transistor no longer has the necessary bias to continue conducting, and so becomes a very high resistance. This, in turn, causes a change of current in the opposite direction, which produces an opposite-polarity signal from the transformer and so speeds up the cutoff action.

When current flow is completely cut off through the transistor, a relatively large amount of energy is stored in the magnetic field of the transformer and has to go someplace.

When the field collapses, it induces a signal in the base-emitter winding, which provides turn-on bias for the transistor, and the cycle begins again.

Thus each cycle in the output waveform consists of a period during which current is building up, a period when current flow is zero, and a sudden transfer between these two conditions. Unlike the ac we looked at in Chapter 2, the output of this oscillator is not at all like a sine wave in its form.

Figure 7-3 shows approximately what the output of this circuit looks like when viewed on an oscilloscope. When supply voltage changes, the slope of the rising part of the waveform changes (higher voltages make it rise faster), and this is why the output tone depends upon the voltage.

This type of circuit is known as a "blocking oscillator" because the oscillator blocks its own action; they are widely used in industrial electronics to produce narrow pulses, but commercial applications use much more complicated circuits to achieve their results.

BATTERY-SAVING AMPLIFIER

One of the big differences between transistors and vacuum tubes, aside from the question of power consumption, is that transistors come in two polarities (pnp and npn), while tubes work only one way (corresponding to npn). This may not seem to be such a striking difference to those of us who never worked extensively with vacuum tubes, but it makes possible many circuits which could never be constructed with only one polarity of device. This project is one of those.

It's called a battery-saving amplifier because it requires only one 3V battery and draws very little current.

The **complementary symmetry** circuit, made possible by using one pnp and one npn transistor, is directly responsible for the single-battery capability, and also for the low parts count. In addition to the 3V battery (which can consist of two penlight cells), you'll need one npn transistor, one pnp, one 10K **audio** taper volume control, one 330K half-watt resistor, one 22-ohm half-watt resistor, and one 1 uF 3V electrolytic capacitor. Any larger capacitor can be substituted, and 20-percent resistor tolerances are permissible.

The general-purpose transistor required for the preceding project can be used here as well. The npn unit can be any available germanium unit.

To listen to the output, you'll also need a high-impedance headphone. The 2000-ohm variety is recommended; low-impedance phones such as stereo units or the small transistor-radio earpieces do not work satisfactorily with this circuit.

The entire circuit is shown in the schematic, Fig. 7-4. You can easily build it on a piece of cardboard by punching holes for the component leads with a pin or needle.

For best results, connections should be soldered. If you're not familiar with soldering techniques, take some time out before starting construction of this project and practice soldering wire scraps together. You'll need a soldering iron and a supply of solder for this; use one of the small "pencil" irons rather than a large tinsmith's soldering iron (and don't even think of trying to work with a blowtorch-driven iron; only an electric soldering iron should be used in electronics). Medium-heat elements are best for working with transistors; low-heat units make it necessary to hold the iron on the joint far too long, causing heat damage to the components of the

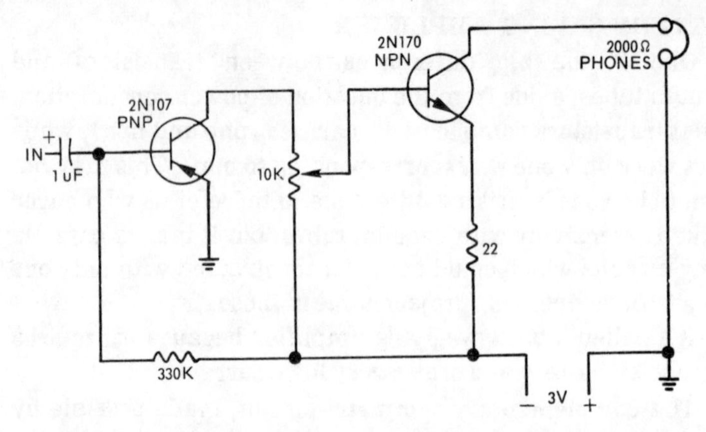

Fig. 7-4. This two-transistor amplifier offers more than enough gain for most practical purposes, while using only two transistors, two resistors, a volume control, and an input capacitor. The input capacitor is not necessary for some signal sources. The 3V battery can be a couple of AA penlight cells, since this circuit draws almost no current.

circuit. A medium-heat element gets the joint hot before the heat has time to travel up component leads and cause damage.

For successful soldering, the wires to be joined must be clean. Scraping them with a knife until they shine is one good way of cleaning them; many experimenters use a scrap of fine sandpaper or emery cloth, folded with the abrasive side on the inside, and pinch this over the wire. By pulling the wire through and twisting it at the same time, the surface is cleaned bright with little effort.

Even though the wires are clean when you start to solder, the heat of the iron will cause them to oxidize unless a flux is used. This is why it's difficult to solder aluminum; the metal oxidizes so rapidly that normal fluxes cannot keep it clean. For electronics, rosin is the only flux to use. The acid-core solder you find in hardware stores is for joining galvanized metal, and leads to nothing but trouble if used in electronics construction.

The solder to use, then, should contain a rosin flux core. Solder comes in several types and sizes, with 50-50, 60-40, and 70-30 being most common. These figures refer to the ratio of tin and lead in the mix. The 70-30 or 60-40 types are best for transistor work as they have the lowest melting points. The best size to get depends upon personal preference, but a "fine" gage is large enough to work with while being small enough to keep from getting huge blobs of molten metal on the iron.

The iron should be up to operating temperature to make a good joint. This is why we do not recommend a gun for your first projects; it's hard to control the head of a soldering gun.

You can tell when the iron is hot enough by trying to melt a bit of solder from the tip of the iron. A hot iron will cause the solder to flow easily into a shiny smooth ball; if the iron is too cold, the solder will not melt. If it's almost, but not quite, hot enough, the solder will break up into a grainy paste; a few seconds later, the grains should flow together into a smooth (and very hot) liquid. If not, the heat element needs replacing with one that generates more heat.

The object in soldering is to cause the molten solder at the joint to dissolve a little of each wire, then freeze into position without disturbance, so that a firm and solid crystal structure results which makes possible a low-resistance path for electric current from one wire to the other. For this to happen, the joint must get hot enough for the solder to become completely liquid, stay that way long enough for a microscopic portion of each wire to dissolve in the liquid metal, and then remain undisturbed while it cools below the solder's freezing point (which is several hundred degrees F).

The most common problem beginners have with soldering is failure to get the joint hot enough. If you take the iron away just as soon as the solder starts to flow, you're not going to make a good joint. The second most common problem is permitting the joint to move while it's cooling, which breaks up the crystal structure. Both these defects result in that nemesis we call a **cold solder joint.**

To make a good joint, start with clean wires and a clean soldering iron. The tip of the iron should be well tinned, and should appear bright and shiny, but should not have any loose blobs of solder hanging onto it. A quick flick of the wrist will shake off any surplus solder, but the resulting blobs of solder in rugs and furniture bring disapproval from homemakers.

Bring the iron in contact with the joint, and heat the joint itself until it becomes hot enough to melt solder. It's all right to have a tiny drop of solder on the iron to help carry the heat from iron to joint, since the heat can flow much more easily through liquid than through solid-to-solid contact, but you should have so little solder in contact with the iron itself that you can easily tell when the joint is hot enough to melt solder.

When this degree of heat is reached (which should take only a few seconds with the proper type of iron), flow in just enough solder, melting it direct from the spool to the joint without using the iron as an intermediary, to coat the joint smoothly. A good solder joint looks as if it has been painted with silver paint, rather than being buried under shiny concrete. Especially, the edge between solder and soldered material should be a smooth fillet, rather than a sharp break. Any sharp change of contour indicates a **cold joint**, because when it's hot enough, the solder flows out into a smooth film.

As soon as you have enough solder in the joint, remove both the solder spool and the iron, and leave the joint completely alone until the shiny look vanishes and is replaced by a slightly duller appearance. This indicates that the joint has frozen. It's still hot enough to cause painful burns, but you can handle it at this stage without fear of harming the crystal structure.

Although the procedure has taken quite a bit of space to describe in words, it's not actually difficult. Fifteen minutes of serious practice should be adequate to permit the rankest beginner to make good solder joints. It's a skill which is necessary throughout electronics, and handy other places.

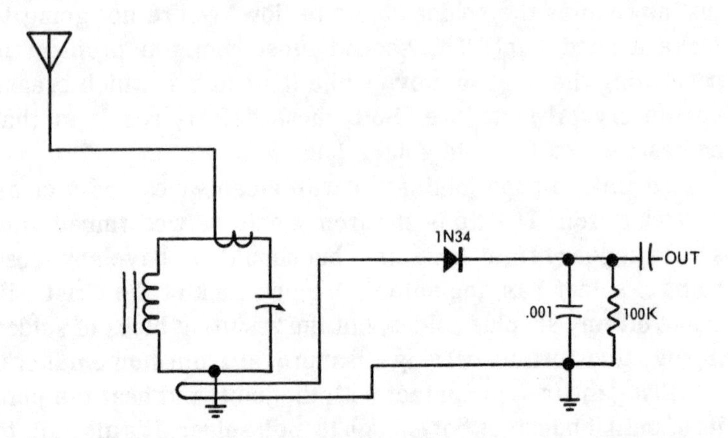

Fig. 7-5. A handful of parts makes a crystal radio set, which can be connected to the battery-saving amplifier. A ferrite loopstick rod and tuning capacitor, both of which may be salvaged from an old transistor radio or purchased inexpensively, tune in the signal. A two-turn link around the loopstick rod couples energy to the diode detector. If you have only one strong station in your vicinity, no tuning arrangement is necessary. Just connect the antenna directly to the 1N34 diode, and listen.

Now that we've taken care of learning how to solder connections, let's get back to the project. After threading the component leads through holes in the cardboard, solder them together following the schematic diagram. Connect the headphones and battery, and the amplifier should be working.

One use for this amplifier is to boost output of a crystal radio set. Figure 7-5 shows a circuit for a crystal set; this circuit will receive stations within a 15- to 25-mile radius, but may not have enough selectivity to separate two strong stations in the same area.

Another use is to provide a "rattle finder" for locating troublesome vibration in an automobile, motorcycle, or boat. For this application, connect a crystal phonograph pickup to the input as shown in Fig. 7-6 (note that the input capacitor is not used in this application), and clamp a wire probe made from a bicycle spoke, piano wire, or welding rod into the pickup's needle socket. When the free end of the probe is touched to any part of any machine (taking care not to get tangled up in moving parts), the pickup converts all vibration of that part to sound, which you can hear in the headphones. A little practice will quickly show you what different kinds of vibration sound like. Most rattles sound the same in the phones as they do normally, but only when the probe is touching the rattling part.

Here's how this circuit works: The npn input transistor acts as a conventional common-emitter amplifier in class A,

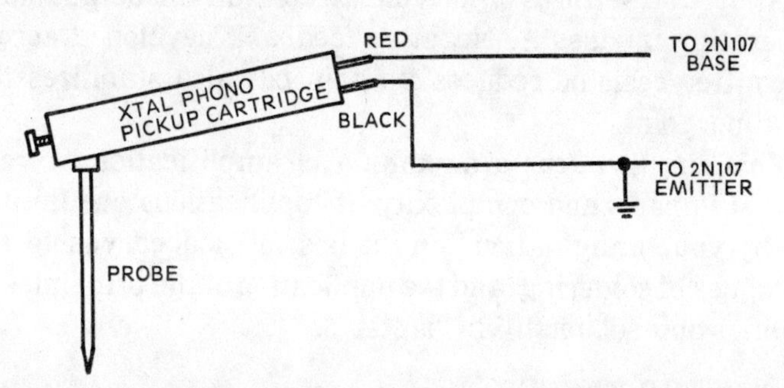

Fig. 7-6. An old crystal phonograph pickup cartridge, with a sharpened bicycle spoke clamped into its needle chuck, turns the battery-saver amplifier into a vibration tracer. The pickup changes vibration into sound, which the amplifier then boosts so you can hear it. The device is handy for tracing rattles in automobiles and other machinery. Note that the input capacitor of the amplifier is not used with this input device, since the crystal cartridge is an open circuit for dc.

with base bias furnished by the 330K resistor. Variations of input signal vary the collector-emitter resistance of this transistor, which in turn causes the voltage drop through the volume control to vary. The collector end of this control, though, is always less positive than the supply end, because the transistor is operating class A and always draws current. This current causes a voltage drop across the resistor, so that it behaves as a battery so far as the pnp transistor is concerned.

The pnp transistor, also, is connected in a common-emitter configuration despite the 22-ohm emitter resistor. The purpose of the 22-ohm resistor is to provide both ac and dc feedback to compensate for the variations in base voltage which result when the volume control is adjusted. Base bias for this device is provided by the voltage drop across that part of the volume control which happens to be in its base-emitter circuit, and varies from zero at minimum volume-control setting up to the full voltage present at the input transistor's collector, when the volume control is at maximum.

When base bias is zero, the output transistor is cut off and conducts no current.

With small values of base bias, such as would result from very low settings of the volume control, the output stage tends to behave as a class B circuit rather than class A, but the distortion is not objectionable because the input signal voltage swing under these conditions is very small.

At normal settings of the volume control, the output stage is operating in class A. Negative feedback developed across the emitter resistor reduces its gain but also stabilizes the operating point.

This circuit offers a maximum of amplification at very low cost in parts and complexity. Its applications are limited only by your imagination—and it has introduced you to the techniques of soldering, and the application of the principles of amplification set forth in Chapter 6.

TRANSISTOR TESTER

Now that we've made use of transistors in a pair of projects, it might be a good idea to have some way of testing those transistors so that we can determine whether any problems are due to bad active devices or to other causes.

While a high-quality tester for vacuum tubes is a costly device, it's a simple matter to build a tester capable of testing all transistors, and get back change from a $10 bill. All you need is a meter, four resistors, two pushbutton switches, a 6V battery, and two test sockets. If you want to make it look professional, add an aluminum box chassis and you're in business.

The circuit for this tester, shown in Fig. 7-7, is based on a design first published a number of years ago in an early edition of the GE transistor manual. We've adapted the circuit to include battery test capabilities.

The tester is built around a 0-3 mA moving-vane **Shurite** meter with a series resistance of 760 ohms. The 680-ohm resistor shown in Fig. 7-7 adds to the meter's internal resistance to produce a total meter resistance of 1440 ohms, which makes it act like a voltmeter with full-scale reading of about 4.33V (3 mA times 1440 ohms). This makes it short-proof. If some other meter must be substituted, change the value of this resistor accordingly, so that the meter gives full-scale deflection with 4.3V applied to meter and resistor together as shown in Fig. 7-8.

The other parts should not be difficult to locate. Both 200K resistors should be 5-percent tolerance to assure accuracy of the finished instrument. The two test sockets can be of any

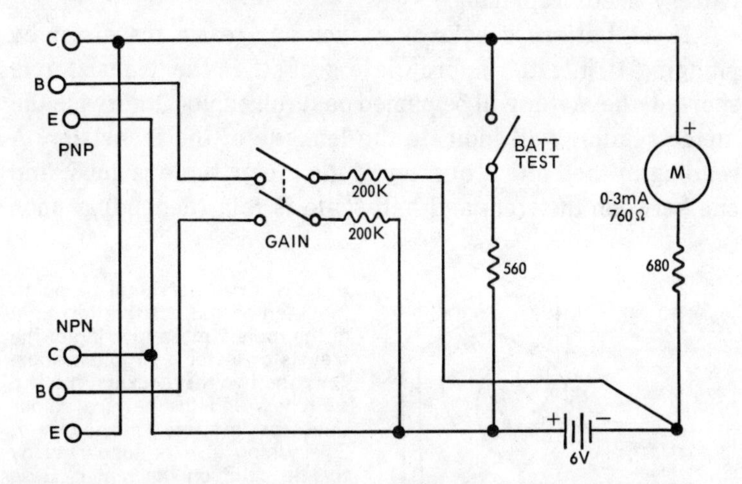

Fig. 7-7. All-purpose transistor tester detects shorts and opens and checks gain for both **pnp** and **npn** transistors of all types. Wiring is not especially complex despite versatility of device. This is all there is to it.

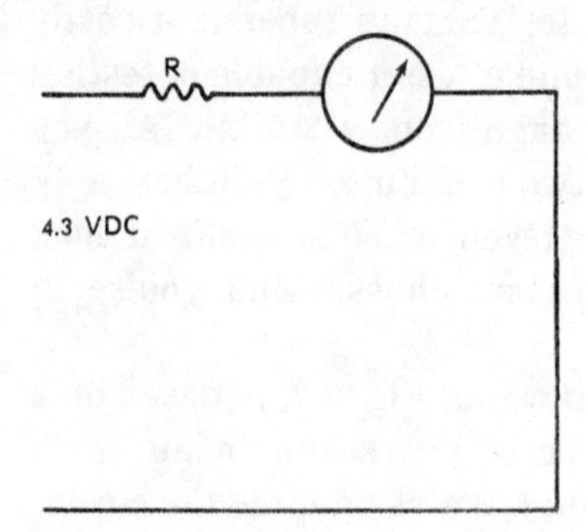

Fig. 7-8. If you have to substitute another meter for the one specified in this project, you'll need to borrow a voltmeter and set up a dc source of 4.3V to find a resistor (R) which will provide a full-scale reading on the meter you're going to use. Then use this resistor in place of the 680-ohm unit specified. You'll also need to change the value of the 560-ohm resistor in the BATT TEST circuit so that it gives full-scale readings with a fresh battery in the tester.

type you prefer; simple leads with alligator clips on them might be best if you anticipate testing a wide variety of transistor types.

Most of the components can be mounted on a terminal strip attached to the meter's binding post as shown in Fig. 7-9. Only the meter itself, the sockets, and the two pushbuttons need be mounted on the front panel. Take care when cutting mounting holes to do a neat job. Aluminum cuts readily with woodworking tools, but it can leave sharp burrs which cause nasty cuts unless you're careful.

Here's how to use the completed tester. The **BATT TEST** button is used only to verify battery condition. When it is pressed the meter should read exactly full scale; if not, the battery needs replacing.

If the battery checks okay, you can test a transistor by plugging it into the appropriate socket. If the transistor is shorted, the meter will be pinned past full scale. Otherwise the meter reading will indicate the leakage of the transistor. A reading in the bottom quarter of the meter scale is good, and one between quarter- and half-scale is fair (depending upon

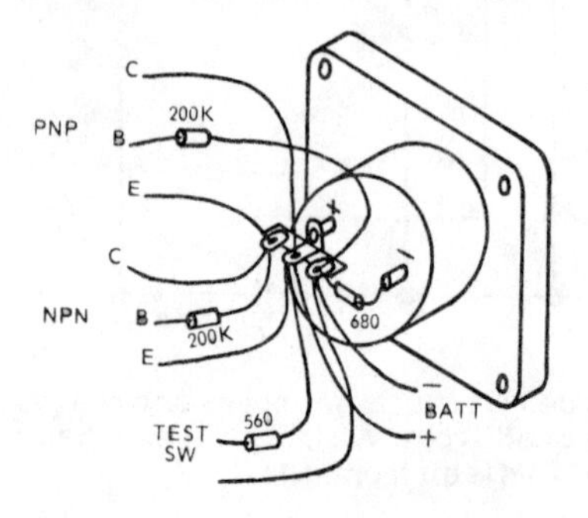

Fig. 7-9. One 3-terminal tie point, mounted on the positive terminal of the meter, makes wiring of the transistor tester a simple matter. Only the two test sockets, the two switches, and the battery holder need be mounted separately. Everything else is supported by the tie point on the meter. Base terminals shown here do not go directly to test socket, but pass through DPST GAIN switch first.

the application for which the transistor is to be used), and anything over half-scale is too leaky for most uses.

If the transistor passes the leakage test, press the **GAIN** button to check its gain. If the meter reading does not change, the transistor is bad. Total absence of any meter reading in both the leakage and gain tests indicates an open transistor. The reading for a good transistor will jump when the **GAIN** button is pressed. The greater the jump, the more gain the transistor has.

Here's how the circuit works: The two sockets take care of the different polarity requirements of pnp and npn transistors. If you use clip leads, only four leads are necessary, because the collector of a pnp and the emitter of an npn connect to the same point, while the emitter of a pnp and the collector of an npn also share the same connection point. Only the base leads are completely different.

When a transistor is connected into the circuit, the meter measures its leakage current directly. This is the current flowing through the emitter-collector circuit with the base disconnected. If the transistor is shorted, the 6V battery drives the 4.3V meter circuit off scale, but the overload is not great enough to damage the meter. The **BATT TEST** button simply connects a 560-ohm resistor in place of the transistor collector-emitter circuit, producing full-scale deflection with a 6V battery. If the meter is changed, the **BATT TEST** resistor may also require changing to make this work out for exactly full scale. Alternatively, if less than full scale is produced by a fresh battery with a different meter, the "good battery" point may be marked and the resistor left alone.

When the **GAIN** button is pressed, base bias is applied through one of the two 200K resistors. With a 6V supply, the base current is approximately 30 uA—the exact value depends on whether the transistor being tested is silicon or germanium, because of differences in the barrier voltage of the two elements. This would cause a 3 mA increase of current for a transistor having a beta value of 100, which would make the meter go to full scale. If beta is only 10, the needle would still jump up by 10 percent of the scale, and this small a beta value is almost unheard of in any transistor which works at all. Thus, any deflection greater than tenth-scale when **GAIN** is pressed indicates that the transistor being tested has gain.

While the circuit of this tester is simple, it does just about everything that any transistor tester available to home experimenters can do. If your transistors consistently indicate off-scale during the gain tests and you would like to get a more accurate idea of their comparative gain, you can use higher resistance values for the 200K units. Be sure to keep these resistors both at the same value, though, if you want pnp readings and npn readings to be comparable to each other.

TRANSISTORIZED VOLTMETER

Now that we've started building test equipment, let's keep it up and put together a transistorized voltmeter. This instrument can be used to measure voltages in all electronic circuits, and offers high input impedance so that it has little "loading" effect on the circuit being measured. Only two transistors, one zero-adjust potentiometer, two load resistors, a battery, a range switch, and calibration resistors are required. The number of required calibration resistors depends upon the number of ranges you want your voltmeter to have.

This circuit will introduce you to the **differential amplifier**; that's a special type of amplifier composed of two conventional amplifiers opposing each other, so that signals applied to the inputs of both with the same polarity cancel out in the output. Only **differential** signals (that is, those having different polarity or phase at the two inputs) appear across the two output terminals. The differential amplifier is widely used in industrial electronics because it automatically eliminates many types of unwanted signal.

Figure 7-10 shows the schematic of this device. The two transistors and the two load resistors can be mounted on cardboard attached to the meter terminals, with the meter, zero-adjust potentiometer, and range switch mounted on a panel.

The meter should be of at least 100 uA sensitivity, because the inexpensive transistors specified do not respond linearly enough for use as an accurate voltmeter over a wider current range.

Values necessary for the calibration resistors cannot be determined until you have put together the basic meter, since they are determined by the amplification of the specific transistors you use. Therefore, you have to put together the basic circuit shown within broken lines in Fig. 7-10, then determine calibration-resistor values by trial and error.

Before attempting to determine resistor values, the meter needle should be zeroed by adjustment of the potentiometer. Once it is zeroed, leave the instrument alone for 15 to 30 minutes to determine whether the zero adjustment drifts. The needle should still be on zero after this time. If not, touch each transistor in turn, and replace the one which did **not** cause the needle to move back toward zero when warmed by the heat of your fingers. Then repeat this zero-drift test. (Some transistors are more sensitive to temperature change than others; this test assures that both your transistors are fairly well matched in temperature sensitivity.)

To select values for the calibration resistors, you'll need a 5-megohm potentiometer, a 1.34V mercury cell, and the use of a reasonably accurate ohmmeter (which need only be borrowed, if you know anyone who has one). If you can't get the use of the ohmmeter, you can get by with a large collection of resistors from which to choose. The idea is to adjust the 5-megohm pot to the required value, then find a fixed resistor of the same value and use it as the calibration resistor.

The lowest scale normally of much use should be 3V full scale. The calibration resistor for this scale should be of such a value that the reading obtained from the 1.34V mercury

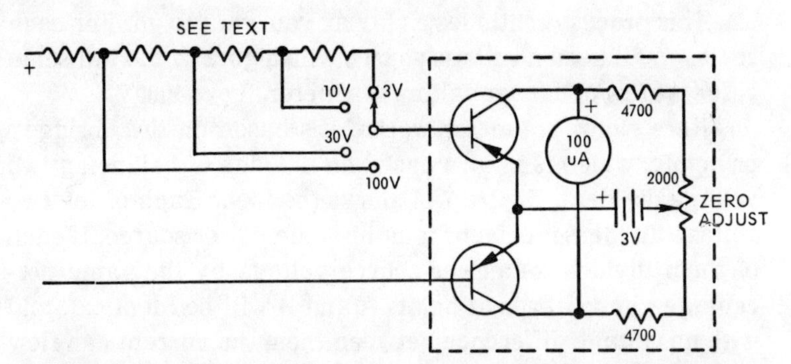

Fig. 7-10. This transistorized voltmeter has higher input impedance than most electronic voltmeter circuits at voltage ranges above 30V. Exact values of calibration resistors will depend upon your transistors, but sensitivity of around 500K to 1 megohm per volt can be anticipated. This corresponds to current gain of 50 to 100 in differential amplifier circuit. Text explains how to choose calibration resistors. Other components are not critical.

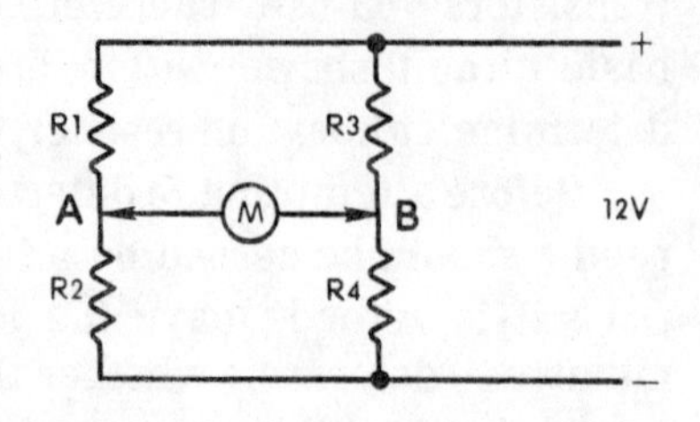

Fig. 7-11. The bridge circuit is widely used. It's based on a pair of voltage dividers. Here, R1 and R2 compose one divider; the other divider comprises R3 and R4. If the ratio R1:R2 is the same as that of R3:R4, the voltage at each junction will be the same (even though R1 and R3 may not have the same value), and so no current can flow through the meter. Changing any one of the resistors will then unbalance the bridge by changing one of the division ratios, and another resistor will have to be changed to modify the other ratio accordingly, to return the bridge to balance. In this project, we do not rebalance the bridge, but instead measure the amount of imbalance by reading the meter.

cell is just below half-scale (the exact ratio would be 1.34 : 3). Once you have found the proper value to achieve this result and connected it into the circuit with the range switch, you can use the 0-3V scale just calibrated to help set up the next higher scale (0-10V).

One of the simpler ways to do this calibration is to use a 10K potentiometer connected across an auto battery or other power source providing more than 3V, and adjust the pot until the meter indicates exactly 3V on the 0-3V scale. Now switch to the 0-10V scale and find the calibration resistor value which lets the needle indicate 30 percent of full-scale deflection. That's the value for the 0-10 scale, and you are now ready to calibrate the 0-30 scale.

This process continues as far as you want to go. For each scale, you'll need a voltage source which goes to the full-scale value, so few of us ever calibrate a meter above 1000V.

Here's how the meter works. It's based on the "bridge" principle, which is in turn based on the idea of balancing two voltage dividers. Figure 7-11 shows the idea. Each of the two voltage dividers is driven from the same power source. If each of them divides the power source voltage by the same percentage, the voltages at points A and B will be identical, and with no voltage difference between them, no current can flow between these points.

If, on the other hand, point A is a fraction of a volt more positive than point B, then current can flow from A to B. Similarly, if B is more positive than A, current flows in the reverse direction from B to A.

The actual values of the individual resistors in the bridge have very little effect on the action; what counts is the ratios which establish the voltage division. If the ratios are equal, no current can flow; if not, current flows toward the lower-voltage point.

In the differential amplifier, the active elements take the place of the lower resistors in the two dividers. The zero-adjust potentiometer varies the relative values of the upper resistors to compensate for differences in the resistance of the two transistors. When the pot is correctly set, the bridge is balanced and no current flows.

When current is applied differentially to the bases, the resistance of one transistor is decreased and that of the other increases. This unbalances the bridge, and the meter indicates the resulting current flow.

If the current were applied in the same polarity to both bases (returning to the common-emitter connection), both transistors would increase or decrease their resistance together, and the bridge ratio would not be seriously affected.

This circuit is especially useful for dc amplifiers such as the one in this project, because it cancels the effects of temperature changes. Without the differential action, any change of room temperature (or even the heating caused by use of the device) would have the same effect on collector current as would an input signal. Zero drift would make the meter unreliable. The effects of temperature, though, attack both transistors simultaneously, and so are canceled as **common-mode** signals. Only the differential desired input signal gets through to have any effect on the meter reading.

The action works properly only over a relatively small range of temperatures or currents, because of the same curvature of device characteristics which introduces distortion into more conventional amplifiers. That's why a relatively sensitive meter is necessary, and it's also why the zero-adjust control is placed out in plain view where it can be set up just before taking a reading. The range can be extended, but it requires much more complex circuitry to do so.

VACUUM-TUBE POWER SUPPLY

All of our projects so far have used transistors. These devices are by far the most practical for home experimenters,

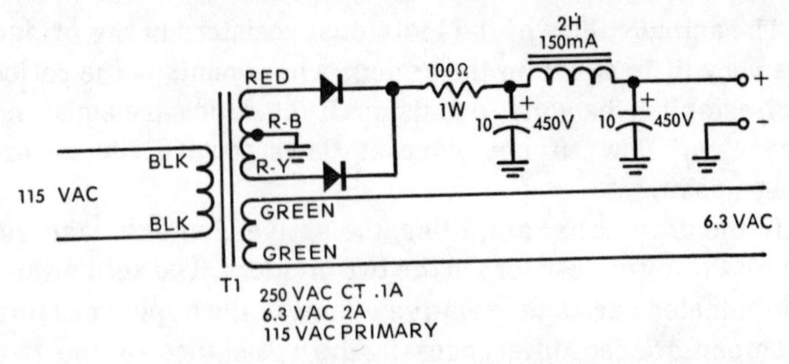

Fig. 7-12. This utility power supply provides adequate high-voltage dc and low-voltage ac to power a vacuum-tube project using up to three tubes of average power ratings. Only critical point in the wiring is the polarity of the two diodes. If top-hat diodes are used, "brim" of hat is normally the cathode end of the diode (arrow in symbol points to cathode). Other types usually have polarity stamped on side of diode. Also take care that filter capacitors are polarized correctly.

because they require much less power. This project helps prove this statement; it doesn't do anything at all, in itself, but provides you with a power source capable of furnishing operating power to many vacuum-tube circuits. That is, this entire project takes the place of the 3V battery we've been using with transistors!

You'll need an aluminum chassis, the transformer specified in the schematic, two silicon diodes, two filter capacitors of at least 10 uF each, and the filter choke specified in the parts list on the schematic. Don't expect much change back from a $20 bill if you purchase all this new. Small parts required include a terminal strip for output, a fuseholder and fuse, and an ac line cord. You may want to include the optional switch and pilot light, to tell when the power supply is on.

Figure 7-12 shows the wiring, and Fig. 7-13 indicates approximate layout of the components on the aluminum chassis. You'll find construction goes easiest if you first drill all required holes in the chassis, then fit in grommets as required to keep the metal from rubbing through the insulation of wires passing through, and finally mount the transformer and choke in place. Now you can install the smaller components by their leads, using tie points as necessary to provide insulated places for loose wires to connect.

Be especially careful when connecting the diodes in the circuit; if they're put in backward, the filter capacitors will be destroyed when you apply power. If you have any doubts, leave the filter capacitors disconnected, then complete the other circuit wiring and apply power. Being careful to avoid getting shocked, check the output voltage with a voltmeter. If the filter choke indicates positive voltage with respect to the ground terminal, the diodes are in properly. If negative voltage is indicated, the diodes are backward. If no voltage is indicated, one diode is right and one wrong; get two new diodes, put them both in with the same polarity, and try again. Only when you get a positive reading should the capacitors be connected. Be sure the power is off before attempting to connect them.

Here's how this power supply works: The transformer steps the line voltage down to 6.3V to provide a low-voltage source for the filament of the vacuum tube, and up to a high value to produce the plate supply's raw input. The high-voltage winding is centertapped, giving us in effect two high-voltage windings in opposite phase when we ground the centertap.

The two diodes rectify these two opposite phases, and each provides positive-going half-cycles to the common point. Since the phases are opposed to each other, the positive-going half

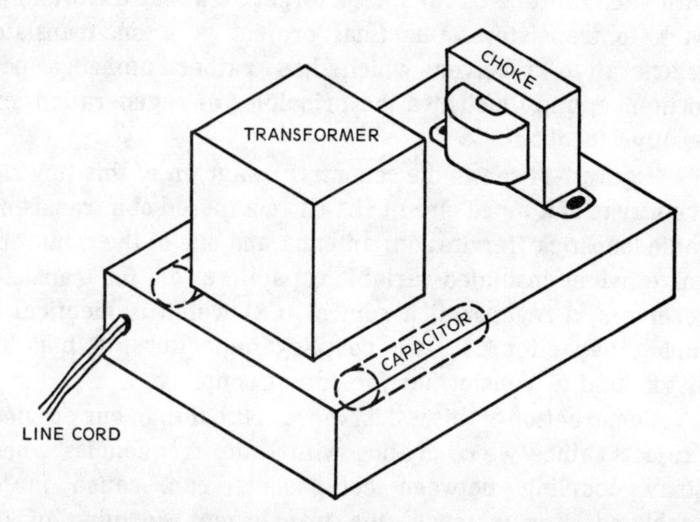

Fig. 7-13. Follow this general arrangement when placing parts on the chassis. Exact dimensions and locations of mounting holes will depend upon the components you use. Filter capacitors can be concealed beneath the chassis. Use tie point strips to provide mechanical rigidity for both leads of each capacitor, and support for the rectifier diodes which should also be mounted beneath the chassis.

cycle of one fills in for the missing half-cycle of the other, and we make use of the full waveform of the input ac. For this reason, the circuit is known as a **full-wave** rectifier.

The 100-ohm resistor limits peak charging current into the capacitors, and so protects the diodes when the supply is first turned on. Once the capacitors are charged, this resistor has little effect.

After power is applied, the capacitors charge in a matter of a few milliseconds, and then serve to smooth the full-wave pulsating dc into an approximately steady level. The filter choke aids in this smoothing action, by opposing any change in current flow.

Output voltage of this supply varies. It's at its highest value with no load, and then rises to approximately 1.5 times the ac value across one-half the transformer's high-voltage winding. Under maximum load, this falls steadily to about equal the transformer's rating.

REGENERATIVE RECEIVER

Now that we've seen how much more complex it is to work with vacuum-tube circuits, let's forget all about them and go back to transistors. Our final project is a one-transistor regenerative receiver which has rather amazing performance, and illustrates the principles of regeneration and positive feedback.

Figure 7-14 shows the schematic diagram of this tiny rig. It consists of a tuned circuit (L1-C1) composed of a transistor-radio loopstick (ferrite rod) antenna and one of the miniature polyethylene-insulated variable capacitors sold for transistor receivers, a regeneration control (C3) which is identical to tuning capacitor C1, two coupling capacitors, a bias resistor, and a transformer for output coupling.

Construction of this is a bit more critical than our previous projects, since we're dealing with radio frequencies where stray coupling between components can cause major problems. For instance, the transformer coupling of the output is a must, because the collector circuit has rf as well as audio signals in it, and if the rf signals are allowed to flow in headphone leads the regeneration control will not work reliably.

Similarly, both variable capacitors must be mounted on insulating material, but their shafts should protrude through a conducting plate which serves to shield them from the added capacitance introduced by your body when you touch the adjustments.

You'll have to wind the tickler coil on the antenna rod. As in our first project, if things don't work right at first, reverse the connections to the tickler winding to be sure you have proper phasing.

Start the project by mounting all the components, and connect them together as shown in the schematic.

When everything is hooked together, double-check to be sure all connections are correct, and hook an antenna to the input terminal. A short length of wire can be used, but the more elaborate your antenna system, the more fun you can have.

Now apply power, and adjust regeneration control C3 to minimum capacitance. If a howl sounds from the earphones, increase the capacitance of C3 until the howl just disappears. If you do not hear a rushing noise, remove power and try reversing the leads to the tickler coil.

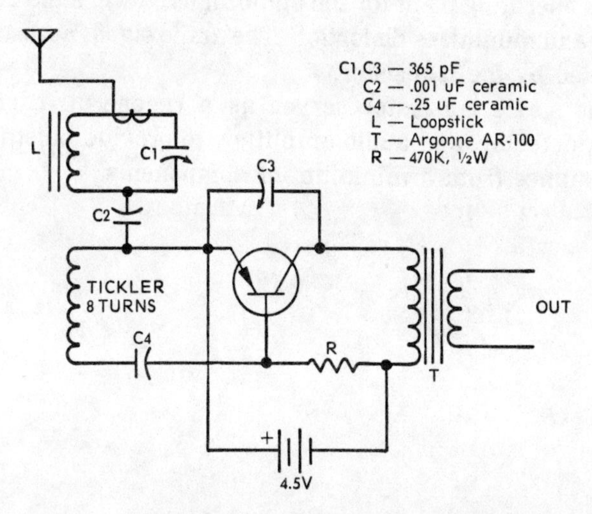

Fig. 7-14. This regenerative receiver provides a maximum of performance with a minimum of parts. Any PNP transistor can be used; a simple germanium type such as the CK722 or 2N107 is recommended. The antenna connects to the loopstick circuit by means of several turns of wire wrapped around the lead from C1 to L1. This "gimmick" capacitor provides ample coupling. See text for adjustment and operating instructions.

When you get the rushing noise in the phones, tune around with C1. As you approach each station, you should hear a whistle. Increase the capacitance of C3 until the whistling noise just goes away. You'll find that the proper adjustment of C3 is different for each station, as it depends to some degree on signal strength. You'll also find that as C3 is closer and closer to the critical point at which the whistle begins, the little receiver is more and more selective.

Here's how it works: Incoming radio signals are coupled into the base-emitter circuit through the tickler coil by transformer action. Capacitive feedback through the base-collector junction permits regeneration, but capacitor C3 from collector to emitter provides a bypass for rf energy in the collector circuit, so that as capacitance of C3 increases, the feedback for regeneration is reduced. This permits C3 to control the amount of regeneration. Resistor R1 provides base bias, which is too little for class A operation; so the transistor operates class B. The nonlinearity introduced by class B operation permits demodulation of the rf signal, resulting in an audio signal across capacitor C2. At the same time, a dc voltage is developed across C2, which helps pull the transistor's operating point for the audio signal toward the class A region and minimizes distortion. The audio signal across C2 is amplified by the transistor.

The single transistor serves as a regenerative rf amplifier, detector, and audio amplifier, to provide outstanding performance from a minimum of components.

Index

Index

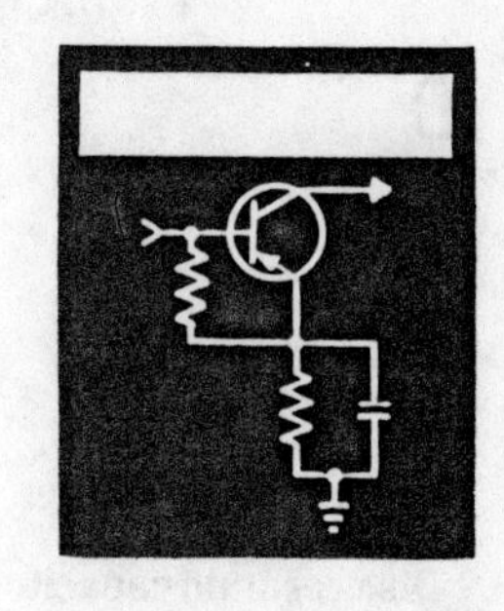

A

B

C

V

W